Bathymetric Navigation and Charting

UNITED STATES NAVAL INSTITUTE
SERIES IN OCEANOGRAPHY

OPTICAL PROPERTIES OF THE SEA
  Jerome Williams

BATHYMETRIC NAVIGATION AND CHARTING
  Philip M. Cohen

Bathymetric Navigation and Charting

PHILIP M. COHEN

UNITED STATES NAVAL INSTITUTE

ANNAPOLIS, MARYLAND

Copyright © 1970

by the United States Naval Institute

Annapolis, Maryland

All rights reserved.

This book, or any part thereof,
may not be reproduced
without written permission from the publisher.

Library of Congress Catalogue Card Number 79–129880

ISBN 0–87021–094–7

Printed in the United States of America

TO
MY WIFE, JOAN

Contents

	FOREWORD	ix
	PREFACE	xi
	INTRODUCTION	3
CHAPTER I:	Soundings	9
CHAPTER II:	The Echo Sounder	17
CHAPTER III:	Signposts on the Ocean Bottom	35
CHAPTER IV:	Charts and Charting	51
CHAPTER V:	Determination and Recovery of Position	75
CHAPTER VI:	Computer Applications and Bathymetry	93
CHAPTER VII:	Needs and Future Developments	107
	APPENDIX	123
	GLOSSARY OF SELECTED TERMS	127
	BIBLIOGRAPHY	133
	INDEX	135

Foreword

When a student recently requested a government agency to send him "a map of the uncharted areas of the Pacific," he received exactly that—a graphic based on extremely sparse and dated information. It is a deplorable and dangerous fact that this situation still exists in vast areas of the ocean. For much of the Pacific, the most recent source of information is the United States Exploring Expedition which Lieutenant Charles Wilkes led in 1838. This is why it is encouraging to see that this book does not limit itself to a description of bathymetric navigation, as important as that is, but that it also discusses the need for surveys and charts for a world ocean of which no area can any longer be considered remote.

In addition to its timely and cogent discussions of surveying and navigation, this book is most relevant to other problems of naval oceanography. The bottom of the ocean is no longer just a place to drop an anchor if one is skillful, or on which to ground a ship if one is not. Increasingly, we are faced with the realization that the sea floor is often the controlling factor in naval operations—from mine warfare to amphibious landings, from antisubmarine warfare to salvage and rescue. But to talk sensibly about the ocean floor, one must first have the basic charts of its features.

Few persons who read this book will disagree with its basic tenets—that the application of automated processes, plus the increased availability of bathymetric data, offers promising and unique approaches to solving deep-ocean positioning problems is beyond speculation. It is equally clear that the assessment of charting requirements must not be considered in light of current needs alone, but rather in the context of the nation's future role and missions in the oceans. I completely agree that while there is imperative need for acquisition of bathymetric data at a rate far greater than present programs allow, more ships and personnel alone will not solve the problem. A coordinated look at specific user needs is long overdue. Then we can begin translating these needs into realistic programs, developing imaginative instruments and methodologies, and working toward a conceptual and common-use approach to data handling. The Navy is, in fact, proceeding along these lines and has made substantial progress towards achieving several of these objectives, but much still remains to be done.

Both in descriptions of particular programs and priority of endeavor proposed, I might disagree with some of the author's views. This in no way detracts from the fundamental competence, impact, and timeliness of the work. Further, in this era of expanding technology, it would be both surprising and disappointing if some of the developments discussed were not soon outmoded. It is strange indeed that there have been only two really significant developments in the practice of taking soundings—the lead line in the mists of antiquity and the echo sounder in the early 1920s. Improvements are surely overdue.

T. K. Treadwell, Jr.
National Council on Marine Resources
and Engineering Development

Preface

One of the consequences of newly increased interest in oceanography is the realization that comparatively little is known of bottom topography in deep-ocean areas. This recognition has been given dramatic emphasis by several deep-ocean disasters in recent years and by the revelation of the scarcity of topographic information associated with these events. It is rather widely accepted that more of this information is necessary for a variety of ocean endeavors, and current plans for surveys to correct this deficiency stand a chance for implementation in some form. Previous efforts had been allowed to wither because of a lack of interest combined with limitations in resources. This was so despite periodic pleas for accurate topographic maps as requisites for studies that would lead to understanding of oceanic processes. The plea began with the report, *Oceanography 1960–1970* prepared by the National Academy of Sciences Committee on Oceanography (NASCO) in 1959. More recently, the Panel on Oceanography of the President's Science Advisory Committee stated, in 1966, that "Topographic surveying is hampered by strict adherence to international conventions developed at a time when the technology was more primitive than it is today." This report assigned immediate priority to offshore areas of the continental shelves, but its comments concerning the gap between traditional concepts and modern technological capabilities are pertinent to deep-ocean (bathymetric) surveys as well. In January 1969, the report of the Commission on Marine Science, Engineering, and Resources, entitled *Our Nation and the Sea*, called for the start of a systematic ocean-mapping program to acquire bathymetric and geophysical data of the United States continental shelf areas to 2,500 meters.

The first attempt to conduct topographic surveys on a systematic basis was made by the U.S. Navy just before World War II, although this was limited to immediate offshore areas in the Caribbean. World War II ended that effort abruptly. The Navy could not afterwards resume the program because of worldwide defense commitments. Project SEAMAP was initiated by the Coast and Geodetic Survey in 1961 to obtain, in part, topographic maps of deep-ocean areas, but full implementation was an early casualty of limitations in resources of men, money, and ships. Future programs of this kind must contend with similar restrictive factors—present techniques are outmoded; the expense is very large; given programs will necessarily have higher priority than others. Should these difficulties be overcome, the effort would still vie with other worthwhile programs for limited funds. This situation redefines the available alternatives, but the ultimate objective remains. More must be known about the deep-ocean areas, and continuation of efforts at the present pace is wholly inadequate. Therefore, we must look to new technologies.

When this book was in its early stages, the application of bathymetric features to positioning was held as its chief purpose. The decision to include discussions on survey needs, and the relation of the subject to matters other than the navigational aspect of positioning alone, came from the considerations described above. It has been attempted to present the contents in an interesting way and free from extraneous technical detail. The result may not be wholly satisfactory to those whose interests are more technical in nature. No one is likely to mistake this book for a comprehensive or definitive work on navigation, echo-sounding principles, survey techniques, or chart construction. Nor can it be thought of as an exhaustive review of computer applications to these disciplines. It is presented in the spirit of bringing the subjects to the attention of a wide spectrum of interested persons. It would be gratifying if the contents came to be regarded as a point of departure for the many specialized studies and works that could properly be devoted to the overall subject matter.

Several individuals convinced me that a book on bathymetry would make a definite contribution to navigation, charting, and oceanography. The late Dr. H. H. Hess of Princeton University, in particular, expressed his belief in the need for such a work and his confidence that I was the person to undertake it. It would be correct to state that he provided sorely needed encouragement.

Dr. Harris B. Stewart, Jr., of ESSA's Atlantic Oceanographic Laboratories, and Captain M. M. Macomber and Dr. J. Brackett Hersey of the U.S. Naval Oceanographic Office, reviewed the book in manuscript form at various stages. Their recommendations were valuable in formulating the final book plan. Dr. Stewart's suggestions resulted in major changes to the text; the debt owed him is large. The bathymetric and general library facilities of the U.S. Naval Oceanographic Office were made available by Captain Macomber and his predecessor, Captain V. A. Moitoret (Retired). Roger C. Taylor of the International Marine Publishing Company, Camden, Maine, generated the impetus behind the inclusion of certain portions of the book and rebuked me on the verbiage sometimes encountered in the early drafts. My wife, Joan, typed and retyped the manuscript more times than she cares to remember.

The views and opinions expressed in the chapter on future developments are personal ones, as are those which may be implied elsewhere in this book. In particular, these views are not necessarily those of any government organization.

PHILIP M. COHEN

July 1970
Silver Spring, Maryland

Bathymetric Navigation
and Charting

Introduction

Bathymetry and bathymetric navigation

Bathymetry concerns the measurement and charting of ocean depths. Bathymetric navigation may be defined as the art of establishing a position geographically in ocean areas with respect to known positions of geological features of the ocean bottom.

Available techniques whereby precise position can be determined and recovered are in many instances not new and, in certain cases, represent mere elaboration and sophistication of methods long in use. What distinguishes their utility today is increased availability of the prime reference source, the bathymetric chart, plus beginning exploitation of computers and computer-oriented systems for bathymetric data collection, reduction, display, and correlation.

The importance of navigation through use of environmental data, such as shape of the ocean bottom, is emphasized by military and other considerations. Such navigation would be even more important if electronic or satellite means should fail or be destroyed. If we ignore for the moment the fact that there is a deficiency in adequate knowledge of the shape of the ocean floor (a serious deficiency to the validity of bathymetric navigation, but one being overcome, albeit at too slow a pace), the most serious drawback to the concept is the average navigator's unfamiliarity with the ways in which environmental characteristics can be exploited. Such characteristics as gravimetric and magnetic anomalies can also be mapped and correlated geographically, but they cannot be applied so readily to positioning since they show variations with time or cannot be measured to required accuracies. Subsurface structures, heat flow, elasticity, and other properties lack geographic relationships necessary to determine accurate position.

Is there real need for a system of navigation based on bathymetric features? Electronic means of navigation covering the major sea routes are in use throughout the world, and satellite methods are now operational. Navigators still have the stars and the sun. Perhaps the navigator off a rocky coast who has not been able to obtain either a loran or a celestial fix for some time would disagree that no further aids are needed, yet it can be argued that, of the matters besetting a commanding officer or a shipmaster today, navigational difficulties are not the most pressing. Probably fire at sea and collisions are of greater concern.

The proponents of navigation by bathymetry might argue as follows. The method uses fixed signposts on the ocean bottom; it is virtually impervious to electronic jamming; it cannot be destroyed as may a navigational satellite; it is worldwide in application, independent of weather or transmission phenomena, and easily learned and applied. Further, bathymetric navigation may be used as an adjunct to any other system. One of the age-old axioms of the navigator is to use every aid available to him.

When man sets forth on the seas, very frequently he must know exactly where he is. Any list of demands for more accurate positioning techniques must include marine salvage, mineral exploitation including offshore oil and deep-ocean mining, implantation on the sea floor of beacons (which themselves may be used for positioning), hydrophones and other gear, and location and utilization of fishing beds. The search for the missing Air Force H-bomb off the Spanish coast at Palomares is an appropriate example of the need for precise positioning. Exact location of the physical features on the ocean bottom is one of the factors which must be known for many kinds of oceanographic studies.

The future may present the problem of parcelling underwater land for commercial enterprises, recreational purposes, or farming. How will the surveyor determine the boundaries of such lands? For military application, submerged navigation independent of land or space-based systems is necessary. It is well known that some foreign ships chart positions off the coast of the United States. Presumably a submarine could position itself over or near a known seamount or other underwater feature before firing a missile. The National Aeronautics and Space Administration (NASA) must position its range instrumentation ships with extreme accuracy for monitoring and guiding space missions, and it is doing so using bathymetric features as an adjunct to other means.

It can be stated, therefore, that a system allowing for more accurate positioning and recovery far at sea would be welcomed by many persons involved in numerous enterprises.

Navigation and charting

Navigation is now becoming less and less of an art, and so close to being a science that the inclination is to forget that it requires many skills. In this regard, Bowditch (*American Practical Navigator*) is quoted:

Human beings who entrust their lives to the skill and knowledge of a navigator are entitled to expect him to be capable of handling any reasonable emergency. When his customary tools or methods are denied him, they have a right to expect him to have the necessary ability to take them safely to their destination, however elementary the knowledge and means available to him.

The wise navigator uses all reliable aids available to him, and seeks to understand their uses and limitations. He learns to evaluate his various aids when he has means for checking their accuracy and reliability, so that he can adequately interpret their indications when his resources are limited. He stores in his mind the fundamental knowledge that may be needed in an emergency. Machines may reflect much of the science of navigation, but only a competent human can practice the art of navigation.*

Navigation implies a requirement for positioning, but not necessarily for positioning of a very high order of accuracy. The location of a ship to the order of several miles suffices for many navigational purposes. Location using bathymetric methods can be used for general navigation, but the term bathymetric navigation implies more *precise* positioning. Requirements for positioning can exist for many reasons, some of them stated previously, including that of navigation if desired.

The ocean environment

The ocean is a most difficult subject for scientific investigation because of the medium in which the investigation is conducted. Although the depths of the ocean are mere wrinkles on the terrestrial sphere, they are sufficient to cause great increases in pressure of water. Special platforms are needed for work at sea in the form of ships, buoys, airplanes, satellites, or submersible vehicles. Temperature above and below the interface is certainly an environmental factor. Instruments must be devised; the corrosive effect of salt water on the workings of sensitive gear must be overcome; surface waters are never placid enough. Even the most modern oceanographic ships can be considered comfortable and spacious only when compared to the battered and beaten ships in service for 25 years and more.

Opacity of the ocean

By far the biggest difficulty to overcome is the ocean's opacity. Absorption and scattering account for the ocean's opacity; *absorption* is the conversion of light energy into heat, and *scattering* is essentially a change in direction of light energy produced by particulate matter in suspension. In the ocean, absorption is large and is related to the wavelength of the light involved; scattering, a more important factor, increases as the number of particles in

---

* *American Practical Navigator, H.O. Pub. No. 9*, United States Government Printing Office (Washington, D.C.: 1962), p. 61.

suspension increases. Turbidity (the amount of foreign material present) is the major deterrent to visibility; erosion, plankton, and mixing are the causes of changes in turbidity. Since water is more dense than air, larger particles remain in suspension in the ocean than in the atmosphere. Therefore, just as easy as it is to map land by man's eye or the eye of a camera from aircraft, it is that difficult to see or chart the ocean bottom. Instead of light energy, sound energy is used for charting or navigating.

Bathymetry: surveying, charting, and navigation

Because of the continuous changes on charts, one might deduce from comparison of undersea topographic charts produced many years ago with those of today that the ocean bottom has become increasingly irregular. This is ascribable less to the irresolute forces of nature than to advent of the *echo sounder*. The present knowledge of bottom terrain features is due primarily to the development and sophistication of the echo sounder. However, advances in echo sounding during the past twenty-five years, such as achievement of transducer directivity, perfection of the conformal array-type unit, and enlargement of the recorder scale, would have taken survey techniques to a limited point only unless there had been attendant improvement in electronic control systems.

Subsequent chapters of this text are devoted to soundings, the echo sounder, the ocean's geological features, chart surveys and production, bathymetric-positioning techniques, computer applications, and present needs and future developments. Interdependency exists among these factors of instruments, charts, and navigation correlation techniques, and much remains to be done to better integrate this relation. Advances in certain techniques have outpaced developments in the others, as in the extreme accuracy of specific depths being frequently superfluous when position is not known to an equal degree of accuracy. Another essential failing is simply the lack of proper reference charts.

In view of navigation by bathymetry, whose principle recognizes the need for correlation of observed geological data with topographic features seen on the chart, some elementary knowledge of submarine geology is required. This subject is covered in chapter three. This appears as a survey of salient geological features which are recognizable on the echo-sounder receiver and on the reference chart.

It is necessary to discuss automation, basic computer technology, and their increasing application to navigation. Because this is fundamental to the understanding of underlying principles, the technical aspects are touched upon but only to the degree necessary.

Lack of a coherent national policy on oceanography has, until recently, delayed systematic acquisition of bottom topographic information. This has resulted in a situation where more is known about some features on the surface of the moon than is known of the ocean bottom on earth. Many pronouncements have been made on the future benefits of more intensive efforts in oceanography. Certainly there may be very large gains to be realized in mining the ocean floor or extracting minerals from sea water. There is little doubt that expanding populations need to find additional food sources, which the sea may offer. It will take longer than is generally realized, however, to develop the technology by which these gains can be achieved economically, perhaps not until the next century. The period during which such technological advances will occur offers an excellent opportunity to obtain maps of the ocean floor which are requisite to the larger goals.

CHAPTER I

Soundings

Early use of soundings

Accounts of early sea voyages make numerous references to soundings as aids to navigation. It is possible to assemble a rather large listing of such references. But, soundings were used as an *aid* to navigation; they were not used as any fundamental *basis* for navigation. Underwater features were used for navigation largely through happenstance. A shipmaster would know of the existence of certain shoals and their location, but sun and stars, sea and wind, compass, instruments, log, and charts were his primary aids for navigating. This was especially true in deep water, which, for practical purposes, was when he was out of sight of land. At times soundings would be used as a check, and if depths were seen to shoal rapidly the ship would undoubtedly slow or stop.

In Elizabethan times, soundings were referred to in the *Rutters,* texts of descriptions and directions, the forerunner of our present *Sailing Directions.* But references to soundings can be seen in the records far earlier than the fifteenth century. Sailing directions of the Mediterranean entitled *Periplus* were written by Scylax of Caryanda about 350 B.C. The application of sounding information to navigation occurred together with development and refinement of gear to measure depths, and consideration of one must include the other.

The practice of taking soundings probably began with man's first venture on the waters. The first man who shoved his dugout onto a shoal soon learned that he needed some way to find the depth of the water, preferably as he was moving. Possibly he used a pole at first. The sounding line was developed early; from the *Historia* of the Greek historian Herodotus we learn it was in use in and about the mouth of the Nile in the fourth century B.C.: "When one gets 11 fathoms and ooze on the lead, he is one day's journey from Alexandria." It is interesting that even then the nature of the bottom was considered significant. Many fishermen during that period were skillful in relating one location or another to the taste of particular kinds of bottom sediment. Clay, in particular, is readily distinguished by taste from other sediments, such as sand, organic matter, and silt.

Also surviving from the age of Greece are accounts of soundings that were sometimes unreliable because of the swiftness of currents. This unreliability and unpredictability had double significance; presumably a line was stretched in swift currents and did not make for true vertical readings, the identical problem noted with lead line sounding even now. Currents could have caused banks to shoal and change position, particularly in delta areas where great amounts of suspended material settled as river velocity decreased rapidly, e.g. Nile Cone Delta and Great Yangtze Sand Bank.

An example of the use of a shoal by Phoenicians is related by Strabo, the Greek geographer. A Phoenician ship was being pursued by Roman warships. To avoid capture, the Phoenician captain deliberately drove his ship onto a hidden shoal and the Romans chose not to follow. It is assumed the ship was able to leave the shoal at the next tide. The Carthaginian Admiral Hanno charted the bottom shape of the Niger River on the west coast of Africa by sounding, and the Greek Admiral Nearchus made voyages navigating, in part, by soundings taken off coasts.

The journey of Saint Paul from Caesarea to Rome, as described in the Bible (Acts 27), is frequently cited as an early example of the importance of soundings. During this voyage, the ship encountered a severe storm. For several days the captain and crew could barely keep the ship afloat. Finally, driven uncontrollably before the winds and waves, they found 20 fathoms by sounding. Shortly afterwards they sounded and found 15 fathoms. Although they could see little in any direction, they anchored and managed to hold for the night. The storm abated and in the morning they found themselves in sight of an unknown rocky shore on which the ship otherwise would have been driven.

Modern pilotage and sailing directions

Fifteenth century directions for the circumnavigation of England carried descriptions of shoal waters and their location, and directions for the voyage to Gibraltar were liberal in use of sounding data. Soundings are contained in many sixteenth century books of sailing directions. In these sources no attempts were made to plot the textually described soundings on a chart, although charts often accompanied the texts. The following extract illustrates the kind of information portrayed:

> To enter Falmouth ye shall fynde a rocke in the myddly of the entrynge, leve it on the larborde syde and go towarde east by those and when ye be past it go streight in for the baye is .... large, anker where ye will at 5 or 6

fadoms amyd the baye and if ye will go at the tournynge of the full sea, for there is a banke to passe which ye shall fynde at lowe water at two fadoms and a halfe.

Compare this with an excerpt from current U.S. Navy Sailing Directions*

> Vessels entering Falmouth Harbor by the main channel from a position about 1 mile south-southwestward of St. Anthony Head Lighthouse should steer in northward with the eastern extremity of St. Mawes Castle bearing 004°, which leads between St. Anthony Head and Black Rock Beacon in a least depth of 44 feet. When the beacon comes in line with the conspicuous turret of Pendennis Castle, bearing about 274°, a conspicuous elm tree on Penarrow Point will be in line with a sharp dip in the skyline of the hills behind bearing 334½°. A pillar on the shore at Pennarrow Point should be kept open to the right of the elm tree. This range leads close westward of Castle Light Buoy, between the buoys marking the respective sides of the Narrows, through Carrick Road, and as far as a position westward of the Vilt Light Buoy in Cross Road.
>
> Vessels enter Falmouth Harbor by the western channel should steer northward and pass about midway between Black Rock beacon and Pendennis Point, taking care to avoid the shoal patches south-southwestward of Black Rock. When the beacon comes in range about 110° with St. Anthony Head Lighthouse, she should steer for St. Mawes Castle and when the channel range (335°) is reached, proceed as previously directed.

It can be seen that sailing directions have come down to us almost unchanged, with little essential differences between the *kinds* of information described being noted. One would assume the shoals are more accurately charted and described in modern sailing directions. These also contain information on berthing, supplies, repairs, communications, quarantine regulations, etc. Today the pace of worldwide marine traffic is so great that even the voluminous sailing directions cannot cope with changes in channels, buoys, lights, and other navigational aids. In the United States and some other nations, a *Weekly Notice to Mariners* containing late information, is made available to every ship, and urgent data are even transmitted directly by radio messages.

Lead line and the Kelvin-White sounding machine

Until modern times the lead line was the only effective deep-sea sounding tool. Essentially this was a length of fishing line with a sinker at one end (the lead) and marked in some way at appropriate intervals by knots or colored tags. At some time the lead portion was hollowed to allow tallow to be placed in it, and so a sample of the bottom could be obtained at the same time as the depth measurement. A lead so fitted was "armed." In many cases there was at least as much significance in the kind of bottom as its depth, and

---

* *Southern Coast of England, H.O. Pub. 31*, (Washington, D.C.: Department of the Navy, U.S. Naval Oceanographic Office), p 83.

by examination of the sample for color and consistency navigators had that much more information to help gauge position. Bottom samples were also of significance as this allowed holding power of the ground to be determined for anchoring.

The use of soundings in this fashion was described by Richard Henry Dana in *Two Years Before the Mast,* published in 1840: "The soundings on the American coast are so regular that a navigator knows as well where he has made land, by the soundings, as he would by seeing the land. Black mud is the soundings of Block Island. As you go toward Nantucket, it changes to a dark sand; then, sand and white shells; and on George's Banks, white sand; and so on."

In the mid-1800s, piano wire was used to obtain very deep lead line soundings. Because of the forward motion of the ship, currents, and crosscurrents, the wire did not hang plumb; there was no way of determining that the angle of the wire was uniform above and below the surface of the water. Because of this unknown angle, the true vertical depth could not be determined while the ship was underway.

In order to obtain an accurate measurement with the lead, the wire had to be vertical; the ship had either to stop or to slow while the sounding was obtained. This proved to be a major difficulty because a successive line of soundings could not be obtained accurately. A method was needed to measure the depth of water independent of length of line paid out so that the sounding could be obtained while underway. As steamships came into use and as the size of ships increased, investigation into causes of shipwrecks showed conclusively that soundings were not being taken because of the delay caused by stopping or slowing the ship.

This problem was presented to Sir William Thomson (later Lord Kelvin) in 1878 and was eventually solved by development of the Kelvin-White sounding machine. This device was a mechanical winch with stops, holding a reel of piano wire by which the lead was lowered to the bottom. At one end of the lead was attached a glass tube coated on the inside with silver chromate (Ag_2CrO) and sealed at the upper end but open at the lower. As the tube descended through the water, hydrostatic pressure compressed the air and forced the water inside the tube to rise. The chemical reaction with seawater changed the red silver chromate coating to white silver chloride.

Thus, if it were seen from the coating the seawater came halfway up the tube, the total pressure had been two atmospheres or one atmosphere above the pressure at the surface, corresponding to a depth of 33 feet.

This sounding machine allowed a line of soundings to be laid down with about a seven-minute interval between soundings. (It took about one minute to reach bottom at 100 fathoms and four to five minutes to retrieve the lead and make ready for the next mark.)

Data from the Kelvin-White sounding machine were more reliable than soundings previously obtained; and although the entire group might be off the true position, the data retained its value for the line since speed and heading were assumed to have been constant. It was valid, therefore, to compare the data with charted values to yield position—if, that is, it were possible to obtain the charted values with similar positional accuracy. This was often not the case out of sight of land. The Kelvin-White system of sounding was the standard in use on most ships until the 1930's and was not displaced to any extent until the echo sounder was developed.

Early years of sound-wave propagation

The first International Hydrographic Conference was held in London in 1919, when echo sounding using principles of sound wave propagation was emerging from the secrecy imposed by World War I. Five men contributed most to the development of the echo sounder since the 1920s. A German, Alexander Behm, invented the Echo Lot and obtained echo soundings in deep water to 150 meters by cartridge firing and microphone. R. A. Fessenden, in the United States, produced the oscillator with receiver which was used for the first Atlantic echo-sounding test in 1922. Pierre Langevin, in France, was working on piezo-electric effects using quartz and finally produced the first supersonic echo sounder, which was the forerunner of echo-sounding methods for navigation since 1930. The first line of sonic soundings was obtained in April of 1922, when the French Hydrographic Department obtained, with an apparatus developed by Marti, a French engineer, a line across the Mediterranean from Marseilles to Philippeville, Algeria. On 20 June 1922 a profile of the North Atlantic Ocean from Newport, Rhode Island to Gibraltar via Nantucket Light and the Azores, was obtained on the U.S.S. *Stewart* by means of the acoustic sounder Type 12, then recently perfected by Dr. H. C. Hayes of the Naval Research Station in Washington.

Credit for development of the first permanent recorder unit for echo sounding is given to Marti. He employed a recorder consisting of a smoked paper strip across which a pen oscillated, making a mark when the echo was received. Somehow this did not receive immediate acceptance, but is was only a matter of time until the advantages of having a permanent record of the depth profile were seen. Moreover, this record could be obtained automatically; this was a distinct advantage over flashing lights on a round indicator, the system American efforts had been developing, or over listening on headphones for echo return, the system on which the British and French were working.

At first, little attention was given to deep-sea features; indeed, the echo sounder was often secured when ships crossed the 100-fathom curve. The instrument was used primarily as an inshore navigational aid, and within coastal areas it became an invaluble tool. Certainly it revolutionized hydrographic surveying and hydrography as portrayed on charts. A bonus to the hydrographic departments of many nations was the ability of ships using echo sounders to check available charts with their own observations. Shipmasters were, and still are, eager to advise hydrographic organizations when discrepancies occurred. The U.S. Naval Oceanographic Office now receives upwards of 20,000 sounding reports annually, in addition to reports received from naval vessels, which submit such data as a matter of routine.

Continuous profiles and geographic control

The marked staff, the line with a weight, the lead line with its many ingenious applications, and early echo-sounding equipment had the limitation of single-point delineation. The modern echo sounder provides a continuous profile of the ocean bottom traversed by the ship, rather than isolated depth. The same is true of recently developed conformal-array equipment which yields closely spaced profiles of an area under the ship at a right angle to the ship. Both types of equipment eliminate the problem of attempting to depict the shape of a bottom feature from a series of random spot depths poorly positioned geographically.

The standard term *sounding* may be defined as the act of determining the depth of a body of water at any position, or defined as the actual depth measurement from the surface of the water to some bottom feature. Today, for a survey to be accurate it not only must have the depth and profile of the water but also must have good geographic position of the soundings. The ability to obtain continuous depth profiles was being developed during perhaps a 15-year period before truly adequate control systems became available, but continuous depth profile and accurate geographic control may be considered, for present survey standards, to have come about at the same time. Before these developments, it was only possible to estimate bottom topography from a series of spot depths. That the depth values themselves were often highly accurate did little to solve the total survey problem, since more often than not the soundings were obtained without adequate estimate of navigational accuracy. This lack of geographic relation compounded the difficulty of ascertaining bottom shape from spot depths alone, in itself often an impossible task. Considering that even now it is possible to obtain conflicting depth information, the task of early submarine geologists in determining shape and extent of bottom features was formidable.

Figure 1 presents a series of spot depths assumed to have been obtained with good control; the spread and value of the soundings appear compatible. The shape of the feature, in this case the top of a seaknoll, can be determined with some semblance of accuracy. Figure 2 depicts the situation that confronted geologists years ago; then determining the shape was difficult, if not hopeless. Echo-sounding profiles must also show consistency of control and position to be valid. Figure 3 can be properly interpreted; figure 4 cannot, unless the ship's track has been adjusted, although these are extreme cases, to be sure. (The true shape of the feature in figures 1, 2, 3, and 4 is assumed to be that shown by the dashed contours.) Frequently it is necessary to align the ship's track to an adjusted position for the correct geological relationship to hold. This adjustment results from the assumption that the track may have been plotted incorrectly, and replotting makes it a better fit with other tracks. In the presence of control incompatability, as well as other phenomena peculiar to echo sounding, development of the shape of bottom features still necessitates intuition as well as the knowledge and experience of the submarine geologist and bathymetric specialist.

Figure 1. Spot depths over seaknoll obtained with good geographic control. In this illustration as well as in Figures 2, 3, and 4, the true shape of the features is assumed to be that shown by the dashed contours.

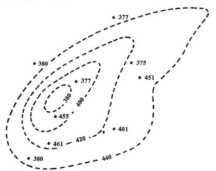

Figure 2. Spot depths over seaknoll obtained with poor geographic integrity.

Continuous echo-sounding profiles have, however, proven exceedingly valuable even in the absence of very accurate control. The gross features of the ocean bottom were discovered during the period when continuous profiles were obtained with only fair electronic control, such as Loran-A, or with no electronic control at all. The Mid-Atlantic Ridge and the major submarine canyons were known to exist before the availability of the reliable control systems now in use. The existence of such features was significant for many reasons, and much scientific work was accomplished even in the absence of extremely good geographic control. However, production of large-scale and even medium-scale maps of the topography of these areas awaited development of accurate control.

As a mechanical device, the permanent recorder type of echo sounder provided an automatic technique only, and was not to be used with blind faith in lieu of skill and craftsmanship. Phase differences, deep scattering

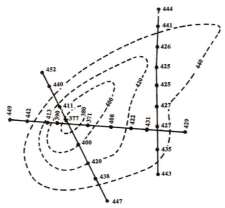

Figure 3. Ship tracks over seaknoll obtained with good geographic control. The shape of the feature can be obtained, although it would not match the actual (assumed) shape indicated by the dashed contours.

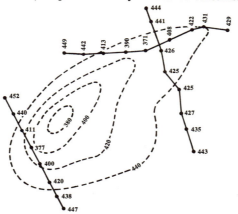

Figure 4. Ship tracks over seaknoll obtained with poor geographic integrity. Track adjustment must be made to a best possible fit, but this is often a difficult interpretive task.

layers, schools of fish, positional discrepancies—all of these caused false soundings to be marked upon the ship's plotting sheet and sometimes upon a published chart. It was simply not good enough to switch on the echo sounder and accept the results obtained while the machine went unattended, and frequent checks and adjustments were required as they are even today.

CHAPTER II

The Echo Sounder

The typical echo sounder consists of two basic units—the *transducer* and the *recorder*. The transducer is both a projector and a hydrophone. A unit designed specifically for reception of sound is termed a *hydrophone*; one designed for transmission is a *projector*. The transducer is reversible and is used both for transmission and reception. The bottom of the ocean acts as a reflector of the transducer's sound waves.

Echo-sounding instruments and their applied techniques on survey operations are being developed at a rapid pace. This is so despite the fact that the echo sounder commonly in use through the 1940's and 1950's, an instrument utilizing a wide-beam pattern from a small transducer at intermediate operating frequency, is still predominant. During the past few years greater emphasis has been placed on increased directivity, or narrower beam widths. This is achieved either by increasing the frequency, by enlarging the transducer, or by a combination of the two, and alternately by arrays of projectors and hydrophones. Normal profile sounding is still accomplished for almost all survey work, but almost ten years of use has demonstrated the feasibility of using equipment that obtains a spread of depth data over a path on either side and under the ship at a right angle to its direction of travel.

The vast majority of ships in the world still have fixed wide-beam echo sounders, and introduction of specialized gear for other than research or survey purposes will undoubtedly be a slow process. This is true of the majority of U.S. Navy ships. Although advanced equipment will have excellent potential for navigational purposes using bathymetric data, techniques are available or can be devised to make maximum use of the standard echo sounder on virtually every ship. Detailed descriptions of the more sophisticated equipment will be given, but stress will be placed on use of the standard echo sounder and its application to bathymetric navigation.

Transducer

The echo sounder's transducer is located in a dome on the hull of the ship at or near the keel, but may be towed to the side or aft of the ship in hydrodynamically shaped holders, or "fish," and connected electrically to the ship. Although towed transducers may be used just under the surface of the water, they are also used at much greater depths. Towed transducers are used more and more, especially for engineering surveys where gear is installed on the bottom, or in search of submerged objects (figure 5). These are sometimes towed very close to the bottom, either positioned by fixed beacons or related spatially to the ship and thence to some other navigational control.

Beam width of sounders

The elapsed time between outgoing signal and returning echo becomes a measure of depth, given the knowledge of the speed of sound through the water medium.

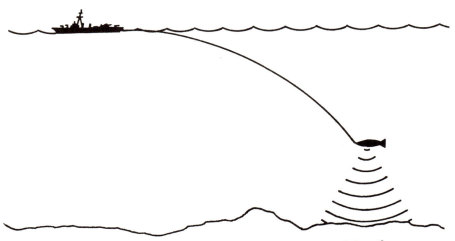

Figure 5. Towed transducer. The "fish" may be towed just below the surface or close to the bottom (drawing is not to scale).

The elements in the transducer housing convert electrical impulses into acoustic energy and, conversely, allow acoustic echoes returning to the transducer to be converted to electrical energy which is visually recorded on graph paper. Recorders usually show the resulting bottom profile at some vertical exaggeration for clarity and measurement. The effective scan range of returning signals is limited to a spread—or beam width—of 50 to 55 degrees in echo-sounding sets commonly in use. This rather broad beam results because the transducer is small in comparison to the transmitted wave length. The projector's echoes reflecting from any surface having an acoustic impedance different from that of water, such as the ocean bottom, fish, the air-water interface, etc., are reconverted by the transducer's hydrophone into electrical energy which is subsequently amplified for presentation on the recorder.

In this type of sounding system, the beam width is a function of diameter of the transducer and operating frequency and can be modified in several ways if desirable. To get narrower beams one must use a transducer that is large in comparison to the wave length of the acoustic signal. Early attempts to decrease beam widths resulted in very large transducers, but some effective balance had to be reached because transducer size can easily get out of hand, some large models reaching 40 inches in diameter. Increases in frequency necessary to obtain directivity were accompanied by greater attenuation of the sound in water, and effective range was thereby reduced. Because transducer size is more or less fixed because of hull installation limitations, directivity in recent years has generally been sought by using a higher operating frequency, although some compromise is required.

The sounding set in use predominantly in the United States uses ammonium dihydrogen phosphate (ADP) as the active elements in a 10-inch diameter rubber-faced transducer. This yields the 50- to 55-degree beam width at 12 Khz operation. The requirement for more accurate maps of the bottom necessitating narrower beam widths caused a 40-inch transducer to be developed yielding approximately a six and one-half degree beam width at −10db at 34 Khz operation. Such a narrow sound cone brought limitations in operating depths as attenuation decreased the effective range of this transducer to between 1,000 and 1,800 fathoms, depending upon both slope and reflectivity of the bottom. Greater range was then obtained by putting a lower frequency through the larger transducer head. Lower frequency (12 Khz) through the 40-inch transducer yielded a 16-degree beam width, but at a much greater range (up to 3,000 fathoms).

The operating range of the standard 12-Khz set is adequate for the greatest ocean depths, close to 6,000 fathoms.

Stabilized, directional, and nondirectional transducers

Highly directional transducers cannot be fixed without stabilization or else their highly narrow beam width would be displaced to the side, ahead, or astern, by the motion of the ship even in a normal sea (figure 6).

Use of directional transducers has proved worthwhile in many instances, but there have been limitations in effectiveness. The range problem has been mentioned, and stabilization is required. The more accurate soundings recorded are sometimes superfluous over flat topography where wide-beam sounders would do as well, and in more rugged topography very good navigational control must be available. In surveying a given area for chart production purposes, complete bottom coverage is frequently desired. Consequently, in relatively shallow areas, where a ship may transverse very closely spaced

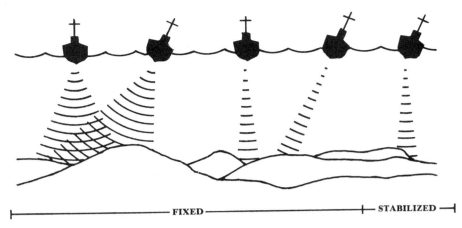

Figure 6. Fixed and stabilized transducers. Irrespective of the beam width of the transducers, those which are fixed to the hull are cocked to the side or fore and aft as the ship itself rolls or pitches. Stabilized transducers always have their energy beamed straight downwards and so receive depths more directly under the ship.

survey lines, the lines obtained are often incompatible because of control problems.

The wider beam characteristics of the standard sounding set are found to be most useful. Often the only means of detecting a feature is by using a wide-beam system in a ship that is not directly over the feature. Because of the cone configuration, the deeper such a feature lies, the greater the horizontal distance at which the ship can locate it. From the bathymetric navigation standpoint, the fact that such a displaced feature is recorded at a depth greater than its true depth is a meaningful clue to its position. The minimum depth recorded by a ship over a seamount will be identical to that shown on the chart only if the ship passes directly over the top, assuming that the charted depth is in fact correct. If the top of the seamount still lies within range of the sound cone, it will be recorded even though the ship is to one side. However, the minimum depth will be recorded deeper than shown on the chart because of the greater oblique distance to the transducer. Within a range of values, the difference in minimum depths (between charted and recorded values) will yield distance horizontally from the ship to projected point of the seamount top at the surface. Finally, in searching given areas for complete coverage, the wider beam yields maximum coverage with fewer survey lines, also an important consideration.

Simultaneous sounding

Techniques making maximum use of the characteristics of both directional and wide-beam transducers have been developed and used for the past 10 to 15 years. One of these techniques is simultaneous operation of directional and wide-beam transducers at different frequencies. This is quite feasible and has been used successfully on many occasions. The ship must have two echo sounder sets, two transducers separated by a suitable distance, and, of course, two recorders (figure 7). The method of operation chosen at given times depends upon depth, topography, control, sea state, and other variables.

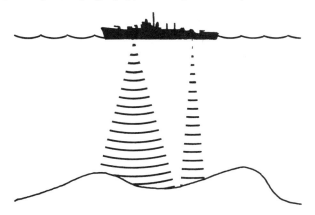

Figure 7. Two echo sounders in simultaneous use from one ship. This scheme is practical when transducers are separated both in distance and by a suitable operating frequency.

As stated before, the elapsed time between outgoing signal and returning echo becomes a measure of depth, given the knowledge of the speed of sound through the water medium. The speed of sound varies from one area to another and is dependent upon elasticity and density, measured from temperature, salinity, and pressure. Almost all U. S. Navy ships use echo sounders calibrated for a standard velocity of sound in sea water of 4,800 feet per second (figure 8).

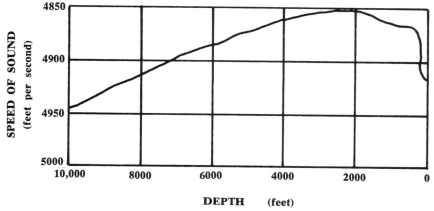

Figure 8. Sound velocity profile shows the typical variation of speed of sound with depth in the ocean. The speed of sound generally increases with the increase in salinity, temperature, and pressure to a certain depth.

Side echoes and hyperbolic shapes

In theory, echoes are returned from the ocean bottom from all points within the area of the sound cone, the area covered being a function of distance from the transducer. In deep water, this area can be quite large (table one). However this large area is represented on the recorder as one single depth value at any given instant. Therefore, as the ship moves through the water, the resultant depth profile is a composite one and may or may not represent the true bottom profile directly beneath the ship's track.

TABLE 1.

Coverage on the bottom by an echo sounder with an assumed 60° beam width.*

Depth (fathoms)	Coverage on Bottom (yards)
100	230
200	460
300	690
400	920
500	1150
1000	2300
1500	3450
2000	4600

* This is given in diameter of an assumed circle. Any ship movement will affect the area covered; appreciable roll or pitch changes the area of coverage on ships with fixed transducers.

First returns to the transducer are sometimes recorded and mask later ones, and in some instances there may be an appreciable delay between the first returns and later groups. The first return is from the bottom nearest the ship, and the successive returns are from other portions of the bottom within the sound cone. These returns of features not directly beneath the ship are called *side echoes*, and over rugged areas are difficult to interpret and distinguish from the true bottom (figure 9). This illustration of returns exemplifies another condition peculiar to wide-beam echo sounders: the tops of peaks or seamounts are, or seem to be, rounded. The recorded shapes are true hyperbolae; each top as seen is a hyperbola.

The significance of this hyperbolic phenomenon is twofold; first, the reasons for appearance of these features and, second, the relation to shape. The hyperbolae develop because a single point of the feature is being recorded for a given time length. As the ship approaches a seamount, its minimum depth will begin to record because it is the point nearest to the transducer even if it is not directly under the ship at that time. That minimum depth is the point nearest the ship when it is approached, as the ship passes directly overhead, and for a period afterwards. The actual shape of the hyperbola on the echogram depends upon beam width of the transducer, the depth of the feature beneath the transducer, and the speed with which the feature is passed.

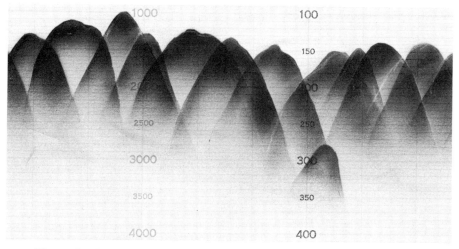

Figure 9. Pattern of side echoes showing typical hyperbolic shapes of peaks. This illustration typifies the difficulties in choosing the true bottom. Only general highs and lows can be found. The condition is caused by wide-beam echo sounders over a complex bottom found in deep waters.

The same hyperbolic phenomenon would apply if the ship passed off to the side, provided the seamount were still within range of the sound cone, and this can sometimes be true even if the top itself were not within range. Since a given point that is recorded forms the hyperbola, there is no relation (except in a gross sense) to true shape of the top with the hyperbola seen. Differently shaped tops record identically if the other factors such as depth and ship's speed are the same.

It is possible to ascertain certain information from the shape of the hyperbola and its depth. Where the bottom is rugged, knowledge of the effect of topography in producing hyperbolae can be useful in interpreting the bottom configuration. Equations of the hyperbola may be used to determine the beam width of the echo sounder, the speed of the ship, and the slope of the bottom. Templates have been constructed that yield this information through a best-fit process. The important point is that in such hyperbolic representations echoes from the top of the seamount are being recorded as the minimum depth as the ship moves towards that position and as it moves away. The minimum depth seen on the echo sounder will be correct only if the ship passed directly over the feature.

Multiple returns

Multiple returns occur in relatively shallow water with a highly reflective bottom, such as sand and gravel. The sound returns from the bottom and is recorded as a measure of the depth; but it continues to reverberate, this time from the air-water interface and to the bottom and then back to the

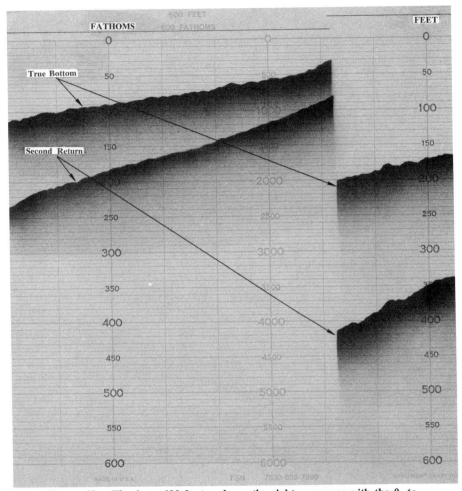

Figure 10. The 0- to 600-foot scale on the right compares with the 0- to 600-fathom scale on the left, both with multiple returns.

transducer. Two or more returns can occur in this way (figure 10). Figure 10 illustrates multiple returns on an echogram showing two scale ranges, 0 to 600 feet and 0 to 600 fathoms.

Deep scattering layer

Where the line of soundings achieved through single depth values by lead line or by early echo sounder often missed significant bathymetric features, the characteristics of the recorded continuous profile sounder often indicated shoals where none existed. In England, France, and the United States, a large number of nonexistent shoals were reported following general introduction of the echo sounder. These reportings were caused largely by the nature of the *deep scattering layer* (DSL), sometimes called the *phantom* or *false bottom*, a suspension of biological matter which was recorded and often obscured or was mistaken for the real bottom. Photophobic marine organisms, the most probable cause of DSL, rise towards the surface at night and sink at dawn (figure 11). Actually the organisms do not scatter the sound, but the many bubbles they produce do. There also may be larger organisms feeding on the smaller ones. Many ships that have reported shoaling in deep water have recorded the deep scattering layer.

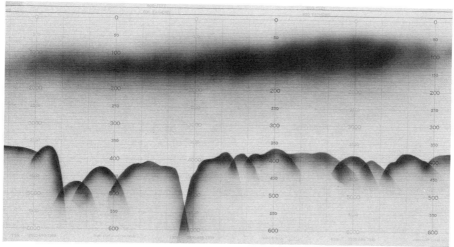

Figure 11. Deep scattering layer (DSL) between the bottom and water surface.

Sharp density discontinuities of temperature, salinity, and pressure also may cause reflection of acoustic sound waves. Echoes may be returned from a boundary of fresh water overlying salt water, for example, although this is uncommon.

In 1948, depth readings of 20-fathoms were reported by the SS *American Scout*, at 46°20′N, 37°21′W due east of Cape Race, Newfoundland. Later that year the SS *American Scientist* reported similar depths in the area, and in the same year the SS *Southland* reported 30 to 90 fathoms. As late as 1964 the SS *Wacosta* reported a depth of 19 fathoms in the same general area. Naval charts previously showed no indications of these relatively shoal depths, but from the safety standpoint the existence of a seamount was indicated upon them. From 1958 to 1966 survey ships found no depths less than about 2,350 fathoms. The surveys conducted by these ships noted significant recording of deep scattering layers, and the true bottom far beneath the layers could barely be distinguished.

It is doubtful that if a seamount of this size existed it could not be located again. If the position of the shoal depths were incorrectly reported originally, this inaccuracy would still allow an extremely large seamount to be displaced only nominally in a horizontal direction when compared to its depth. A typical seamount is very much wider that it is high; the so-called American Scout Seamount would be about 25 nautical miles across at its base. Any displacement caused by positional errors should not have prevented the survey ships from detecting the feature, as the wide-beam echo sounders used would have picked up indications even that much farther from its peak. Further indications (but not necessarily conclusive) that the feature did not exist was failure of magnetometers to show an anomaly, normally correlated with such seamounts. Gravity data would also be meaningful if it had been available.

The feature could still exist in theory, but only if it were an unusually steepsided seamount which does not possess significant magnetic character. This is not likely. The soundings first reported were undoubtedly due to reflection from the deep scattering layer. Because hydrographers are conservative as they are, the seamount was still shown on Navy charts until recently (figure 12).

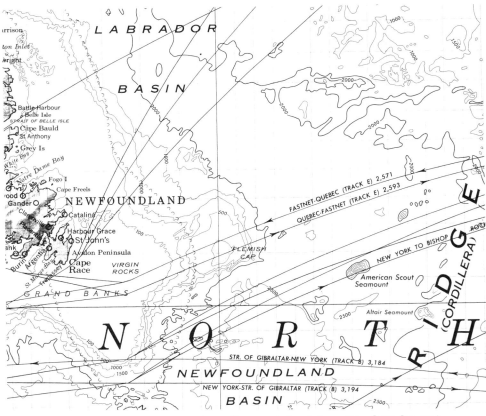

Figure 12. The American Scout Seamount, located on this chart due east of Cape Race, Newfoundland, has been proved not to exist. Indication of the feature on several ships' echo sounders was due to their recording the deep scattering layer in the area.

Range of echo sounder

At first glance it would appear that a key requirement for any system dependent on recognition of underwater features be that such system is able to record all such features at whatever depth encountered. An operating range to 6,000 fathoms is available on many standard sounding sets, and it is correct to state that such range is desired. But the majority of significant geological features which can be recognized from interpreting echo-sounding data are found at depths less than 3,500 or 4,000 fathoms. Many such features are found at much shallower depths, and canyons are found even between 50 and 200 fathoms. Maximum-range echo sounders will, of course, prove more advantageous for all circumstances encountered because the deepest depths known are found in trenches and these trenches may also be indicative of ship's position. Moderate limitation in operating depths would not normally preclude, however, use of a given echo sounder if it is suitable in other respects. Table two lists echo sounders in most common use on U. S. ships other than naval ships, although certain equipment is found on some naval ships.

TABLE 2.
Echo Sounders in Use on U. S. Ships.

Manufacturer and Type	Depth Range
Bendix DR-21	0–150 ft.
RCA CRM-E1A	0–200, 200–400, 400–600 ft.
Raytheon DE-705A	2–120, 120–240 ft.
Raytheon DE-707A	0–300 ft., 0–100 fm.
Raytheon DE-103	2–300, 300–600 fm., 12–300, 300–600 ft.
Raytheon DE-723	1 ft.–250 fm.
RCA Bellatrix	7 ranges 0–280 fm.
RCA Altair	8 ranges 0–560 fm.
RCA Arcturus	8 ranges 0–560 fm.
RCA Castor	8 ranges 0–180 fm.
RCA Deneb	8 ranges 0–2200 fm.
RCA Enif	8 ranges 0–6720 fm.
Bendix DR-7A	0–100 fm.
Bendix DR-9	0–200 ft.
Bendix DR-18	0–20, 20–40, 40–60 fm.
Bendix DR-19	0–60, 60–120, 120–180, 180–240, 240–300 ft.
Bludworth ES-1025	0–60, 60–120, 120–180 ft.
Bludworth ES-120	4–200 ft., 2–200 fm.
Bludworth NK-6	4–200 ft., 2–200 fm.
Submarine Signal Co. NMC-1	0–200, 200–400, 0–2000, 2000–4000 fm.
Submarine Signal Co. NMC-2	0–200, 200–400, 0–2000, 2000–4000 fm.
RCA NMC	0–100, 100–200, 0–2000, 2000–4000 fm.

Recorder of the echo sounder

Perhaps the most significant component of the echo sounder is its recorder. Ocean soundings of extreme accuracy may be obtained by display of flashing lights or upon suitably annotated cathode ray tubes; or a single "ping" can be transmitted and received as a given depth value. These lights, tubes, and pings offer no record with which to work, and the shapes of underwater features are thus impossible to interpret. Flashing lights or display upon scopes of various kinds are therefore unacceptable—besides, there are severe depth limitations to these units, and most show depths of several hundred fathoms only. Such units are often found as ancillary gear to the system and are used to best advantage in shoal waters, entering and leaving harbors, etc.

Bathymetric navigation techniques are applicable to any echo-sounding system that features a visual, permanent recording of the bottom profile under a ship's track. It is the *shape* of the feature that must be obtained; its absolute depth at any point is unimportant. Vertical and horizontal coherence is required, and even if corrections for sound velocity in a given area dictate that the entire feature should be 80 fathoms deeper than calibrated velocity indicates, this would be superfluous for positioning.

The electrosensitive chart paper on the standard recorder is available in rolls of 100-foot lengths and a width of nine and one-half inches. The paper travels across the face of the recorder at one inch per minute at a 600-foot scale, 0.16 inches per minute at a 600-fathom scale, and 0.016 inches per minute at a 6,000-fathom range. Pulse length of the keys also varies with scale range and is 2.5 milliseconds at the 600-foot scale, 15 milliseconds at 600 fathoms, and 150 milliseconds at 6,000 fathoms.

Water depths to 6,000 fathoms are recorded by a stylus on the moving roll of chart paper. The resultant profile of the ocean bottom appears on the recorder at some vertical exaggeration of usually two and one-half times the ship's speed, e.g. about twenty-five to one (25:1) at ten knots. The operator has the option of selecting the 0- to 6,000-fathom range if desired, or ranges of 0 to 600 feet and 0 to 600 fathoms. Variations in paper travel-time in the horizontal, or *x*-axis, occur at different ranges. Continuous automatic repetition of soundings is the usual mode of operation in any of the scale ranges chosen.

Standard Navy echo sounder

The echo sounder shown in figure 13 is found on most U. S. Navy ships, but its features and mode of operation are similar to all sounding systems of this kind. This particular unit bears the military designation AN/UQN–IE. The receiver/transmitter portion and the recorder are mounted on or near the bridge and connected by transmission cable to the transducer on the hull at or near the keel. For special investigations of hydrography, the transducer can be towed free of the hull and, incidentally, free of hull and propeller cavitation. Cavitation is the turbulent growth and collapse of bubbles. Its effect on echo sounding is important because of the noise made by the bubble collapse as the ship steams along. One of the advantages gained by towing a transducer is the elimination (to a large extent) of this effect. The sounding set illustrated weighs less than 400 pounds; some specifications and dimensions are shown in table three. The operating frequency of 12 Khz yields an effective beam width of about 50 to 55 degrees.

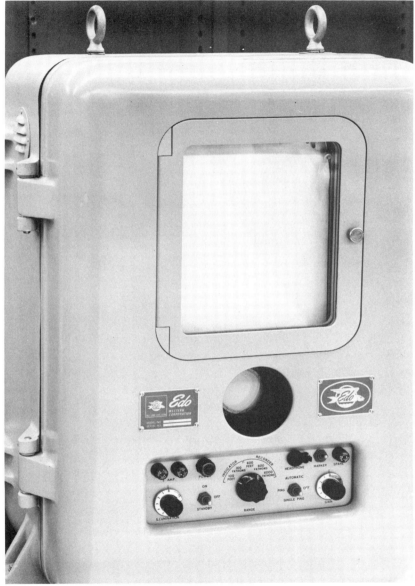

Courtesy of Edo Western Corporation, Salt Lake City, Utah

Figure 13. Standard U.S. Navy sounder-recorder and control panel. The continuous bottom profile is recorded on paper in the window. The lower circle records shallow depths visually by means of a rotating signal.

TABLE 3.

Specifications and dimensions of receiver-transmitter and transducer.

Receiver-transmitter

Overall Dimensions	28¼" high x 22¾" wide x 14" deep
Weight	220 pounds
Recorder Ranges	0–6000 fathoms
	0–600 fathoms
	0–600 feet
Indicator Ranges	0–100 fathoms
	0–100 feet
Paper Speeds	1" per hour (6000 fathoms)
	10" per hour (600 fathoms)
	1" per min. (600 feet)
Paper Length	100 feet
Power Supply	115 volts 60 cycles
	130 watts "Stand by"
	270 watts "On"
Operating Frequency	12 kc.
Illumination	Controllable, red or white
Fix Marker	Push button on control panel

Transducer

Overall Dimensions	14¼" O.D. x 9⅝" high
Weight	150 pounds
Hull Opening Required	10¾" dia. x 2¹⁹⁄₃₂" max. depth
	12¾" mounting bolt hole dia.

The transmitter is capable of delivering initial peak power of at least 1,000 watts for all range settings, average pulse power of at least 600 watts to the transducer at the frequency of the selected range of 600 feet and 600 fathoms, and 250 watts at 6,000 fathoms. The transducer is made watertight by use of an "O" ring seal. A secondary internal seal is capable of withstanding pressures to 1,000 psig (pounds per square inch gauge). The major lobe of the pattern is in the form of a conical beam.

This equipment is designed to operate from a single-phase 60-cycle power source at 115 volts. Calibrated accuracy is within one-half percent of indicated velocity based on a standard of 4,800 fps (feet per second). The 12-Khz operating frequency varies with power source but within a probable effective range of 11.5 and 12.5 Khz.

Enlarged recorder

The primary function of the recorder is to display sonic travel-time intervals in such a fashion that a continuous record of the ocean bottom is produced. The precise and detailed profiles of ocean-bottom topography needed for production of large-scale hydrographic charts are generally unobtainable with the recorder of the standard echo sounding set. Although displayed at vertical exaggeration for measurement purposes and for emphasis of slight topographic irregularities, the recorder of the set shown in figure 13 often is not sufficiently precise to record or read data for chart production. Although the accuracy per individual sounding is ± one fathom in 3000, it does not record nor can it be read with this accuracy when the 0- to 6000-fathom scale is used for recording. The specific requirement for an enlarged-scale recorder was met in 1954 with development of the Precision Depth Recorder (PDR). Since then numerous other models have been developed, and a typical enlarged scale recorder is shown in figure 14.

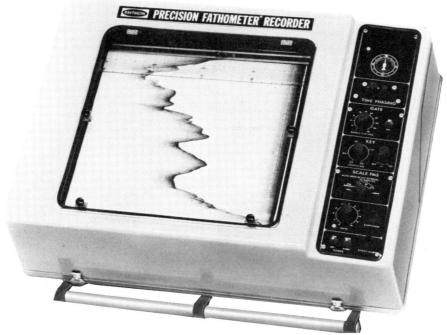

Courtesy of Raytheon Company, Portsmouth, Rhode Island

Figure 14. Large-scale recording device. The paper travel direction is vertical in this model, as is the ensuing bottom profile.

The recorder has several auxiliary functions. It controls the keying (frequency of pulse sent through the water) of the transmitter and the gating (automatic shutting off for a period of time to preclude the recording of unwanted echoes, e.g. DSL) of the receiver, and it provides a time base for correlating depth with ship's position. Other applications include subbottom investigations, scattering layer and sound-propagation studies, instrument positioning, and recording of facsimile-transmitted weather maps when provided with suitable component arrangements.

For its primary function in recording echo-sounding profiles of the ocean bottom, the enlarged recorder has several modes of operation. A small-scale mode of 0 to 4,000 fathoms is available, as well as one of 0 to 400 fathoms, several intermediate scales, and some larger scales of 0 to 20 fathoms. Vertical exaggeration of the bottom profile becomes a function of ship's speed plus the scale chosen, and the width of recording is about 19 inches as compared with about 8.25 inches for the standard recorder. An example of a profile achieved through use of the enlarged-scale recorder is shown in figure 15.

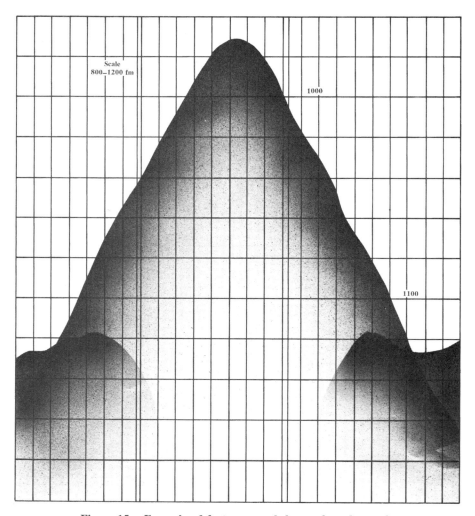

Figure 15. Example of feature recorded on enlarged recorder.

Unless special care is taken in the use and maintenance of any recording system, errors are bound to occur, and some of these can be formidable.

A check on the absolute-depth values must be obtained periodically when working with varying scale phasings, lest a depth of 440 fathoms be read as 840 (or 1,240 or 40, if the 400-fathom scale range is chosen). Slippages in the drive mechanism of the recorder and incorrect zero phasing are common occurrences, and so the recorder must be periodically checked and the adjustments made. The power supply is a factor since this is the basis for the depth annotations. Current is passed through the stylus which remains in constant contact with the electrosensitive paper, and the electrical discharge through the paper causes selective decomposition of the electrosensitive surface. Precision of paper speed cannot be overemphasized, since time is expressed in terms of the width of the recording (x-axis). Radio station WWV provides the time standard used. Station WWV, near Washington, D.C., broadcasts continuous time signals obtained from the U.S. Naval Observatory time service. The signal broadcast by WWV is intended primarily for measurement of time intervals.

Echogram interpretation is discussed in chapter four. This subject could form the basis for extensive writings; indeed, the term *interpretation* suggests that professional hydrographers have achieved no foolproof analysis techniques. Examples can be shown where actual depths, shapes of features, kind of return, reflection conditions of the bottom, etc., are, in fact, subject to individual interpretation. Often this facet of echo sounding becomes an art.

The preceeding description of the typical echo-sounding system and the enlarged scale recorder device is generalized and purposefully devoid of much technical data, but the navigator must have an understanding of such data and an appreciation of the difficulties inherent through use of electrical equipment at sea. Most gear of this type is dependable and rugged; however, constant vigil must be maintained to obtain maximum performance. Technical specifications of the equipment must be thoroughly understood, and competent sonar and electronic technicians must be in charge of the equipment.

CHAPTER III

Signposts on the Ocean Bottom

Basic geology of the oceans

Over the centuries, man's concern with the mystery of what lies beneath the sea has led to tales of lost continents, buried civilizations, and strange beasts. The age of folklore is past, but speculation about the ocean bottom still continues. The vast realm of the seas covers more than 70 percent of the earth; it is sometimes difficult to comprehend that this watery seven-tenths of the world lies relatively unexplored.

Currently, efforts to discern bottom topography are not undertaken so much to discover new major features as to define known structures. It is doubtful in well-traveled trade routes and in areas where surveys have been conducted that new features of great magnitude will be found (although they sometimes are). On the other hand, some features once thought to exist (such as the American Scout Seamount) probably have been charted in error. What is known of the ocean bottom whets the appetite for more detailed and comprehensive knowledge over the entire world.

The features of the ocean bottom, when they have been better mapped, may rival the features of the earth's land surfaces in size and variety. Recent discoveries show that the shapes and scopes of such features are greater than were thought even a few years ago. In contrast to the Grand Canyon, or the Alps, sea bottom terrain is not spectacular, although the Great Meteor Seamount in the northeast Atlantic is half the height of Mount Everest and its summit platform is about the size of Rhode Island. The general nature of underwater forces produces less striking results. Natural forces which shape the lands with which we are all familiar are either absent or less effective in the oceans. Erosion under the seas does not exist in the same way it is known on land; there is neither the wind nor the alternate freezing and thawing; but there are bottom and deep streams, currents, and oceanic water masses. The ocean bottom does not show the sharp effect of continental glaciers which caused relatively permanent topographic changes on land, and it contains no wind and sand-cut cliffs or very sharp crevices. The topography of the ocean bottom, as compared to that of the land, can best be described as *subdued*.

Yet many features that are found in the oceans—ridges, seamounts, escarpments, and trenches—might be considered spectacular enough if they existed above sea level. Many geological features found on land continue under the seas, although their underwater portions may be less severe or extensive. Major earthquake faults extend seaward for hundreds of miles, and mountain ranges in the world earthquake belt are often associated with parallel deeps in the oceans. Some submarine canyons may be associated with young continental rivers, although others appear deepened by submarine slumps or result from submergence of former coastline areas. Further, there are forces at work in the oceans peculiar only to that environment—gullies and discontinuous valleys caused by slumping of water and debris is one—and there is also the buildup of coral reefs and the settling of sediment over centuries. Probably the chief reasons for topographic differences is the nature of erosion in the oceans against that on the continental land masses.

Scientists of the next century will consider the present state of underwater topography as meager, but in contrast to that available a scant 20 years ago a great deal has been learned. While the full potential of surface and subsurface navigation based on underwater topography cannot be reached until more about this topography is known, existing knowledge offers a very good beginning. In theory, the requirements for a given navigational system should exist independent (at the time) of the means to provide necessary input to that system. But this is not always the case—frequently the availability of certain data affords development of new means of exploitation of that data. With respect to bathymetric navigation, development of procedures to use the data will undoubtedly generate requirements for more such data. As this occurs, predictably, newer methods will yet be found for their utilization, and so on. Many years ago a requirement for detailed bathymetric charts at suitable scales and on a worldwide basis could have been stated as a means of affording safe navigation. As such it would have been a futile statement, possibly understood but impossible to achieve. Now, as the need for precise bathymetric navigation grows, increasing knowledge of underwater topography makes fulfillment possible.

Major geological features

Before proceeding with the discussion of bathymetric navigation, it will be necessary to describe the major geological features that make up underwater topography. Definitions will be found in the glossary.

Moving seaward from the shoreline, the first major subdivision of submarine topography is the *continental shelf*. This is a fairly shallow submerged margin surrounding the continents, with a seaward slope of less than five feet per mile. Its depth ranges from 10 or 20 fathoms down to 300 fathoms. The continental shelf may extend seaward from beyond the shoreline for as little as a few miles or hundreds of miles. The continental shelf typically ends where a sharp increase in slope of some 400 feet per mile begins the descent to the deeper waters, which is called the *continental slope.*

On these broad slopes may be found *canyons, seaknolls,* and other features. Characteristics of the slopes depend largely on the nature of the coastline with which they are associated. The base of the slopes may contain *rises, marginal plateaus,* and *trenches.* The continental slopes lead down to the deep submarine features. All these features, with the continental shelves and slopes, constitute what is sometimes called the *continental margin.*

Figure 16. Composite physiographic diagram showing typical underwater features.

Submarine canyons are sometimes found cutting through the continental shelf and slope down to the deep-ocean floor, and others are found in the ocean-basin floor with no apparent land connection.

Notable features of deep-ocean areas are *seamounts, tableknolls, island arcs, rises, hills* and *plains,* major ranges such as the *Mid-Atlantic Ridge,* and deep *trenches.* Figure 16 illustrates those which can often be identified through echo-sounding techniques.

Contouring of geologic features

It is sometimes difficult to relate the information seen on the echo sounder recorder with the geologic feature as seen on the chart. No doubt one reason for this inability to correlate is the paucity of data. Another factor contributing to this difficulty is the interpretive nature of the contouring during the chart construction phase. In the face of insufficient "raw" sounding data, some displaced from correct position and others conflicting in depth (often because of positional errors), it is left largely to the cartographer to determine the bottom configuration by use of contours. With these difficulties it is not always possible to delineate features in a true geomorphological sense. Solely on the basis of "raw" sounding data, it is possible to portray geological features in different ways. This is a far cry from contouring a map, where matched pairs of stereographic photographs allow for concise and unique contour shapes and positions. Comparison of observed data with charted data is difficult even if all charted data were assumed to be correct. The process would be more exact if the chart represented sounding data properly interpreted geologically, and this assumes availability of sufficient sounding data to produce valid contours.

Geological interpretations

The problems of bathymetric presentation can be illustrated by the examples in figures 17 to 20. The tracks of three ships have been plotted as AA', BB' and CC' in each figure. It is desired to contour this section of the bottom at a 100-fathom interval, with the chart at some medium scale, say 1:400,000. The only data that exist in this area (for purpose of this example) are those obtained by the ships whose tracks are shown. Actually, the tracks would not be the straight lines indicated, nor would the fixes at which each depth is marked be of equal spacing. It may be assumed that these have been adjusted. The depths shown are those measured from the echogram at equal time intervals.

Depths in figure 17 do not correlate well where the tracks cross. Tracks BB' and CC' do show some correlation, but a 30-fathom difference exists at the intersection of tracks AA' and CC'. The depths on one of the tracks could have been measured or plotted in error, or one of the tracks could be displaced in position. If these are random tracks of ships other than survey ships, either of these errors is possible. Contouring could be accomplished as in figure 17. Here the individual has chosen to interpret the soundings by drawing almost concentric circles about the highs. At this point not enough is known to say this is incorrect, although it appears unlikely. It is also possible to contour the data presented as seen in figure 18. Here the individual has elongated the highs, but of more importance a definite trend has been established in a NE-SW direction.

Figure 19 presents a different situation entirely. The individual has assumed that track CC' has been plotted incorrectly. Moving the track in a southeasterly direction a certain distance improves the crossings so that good correlation is achieved. This is normally a valid procedure provided there is

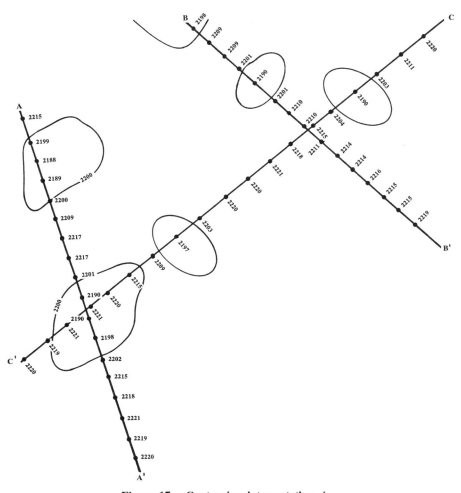

Figure 17. Contouring interpretation A.

good reason to believe such a shift might have occurred. It is often possible to ascertain, from comparison of the ship's sounding log with the navigator's log, that the navigational fixes were taken or plotted incorrectly. And where numerous tracks exist, each crossing the other, a track out of position can sometimes be identified by a process of elimination. If the assumed new position of CC' is now DD', the data can be contoured as shown. If the contours shown in figure 20 are those which actually do exist, i.e., this is the true bottom shape at this contour interval, then it can be seen that figure 18 is the version most nearly correct.

The cartographer who produced the contours in figure 18 presumably reasoned as follows: This area does not exist in a void. The shape of the bottom portrayed in this section is due in the larger geological sense to a kind of topography displayed over many square miles. The contour trend in this overall area is generally NE-SW, and it consists of low, gently rolling topography. It is true that track CC′ appears to be out of position. But where is it? Instead of repositioning track CC′ why not take into consideration its lack of correlation with AA′, and to some extent attempt to rectify the crossing at BB′?

Unfortunately, even such reasoning may sometimes result in incorrect portrayal, despite the validity of the analysis based on the data on hand. The truth is that in areas where no detailed geographic control is possible, normal celestial fixes do not provide the accuracies necessary to achieve an accurate picture of the bottom on any but small-scale bathymetric charts for gross portrayal only. Loran-A control is not good enough, and the example shown would have its difficulties multiplied manyfold in complex topography.

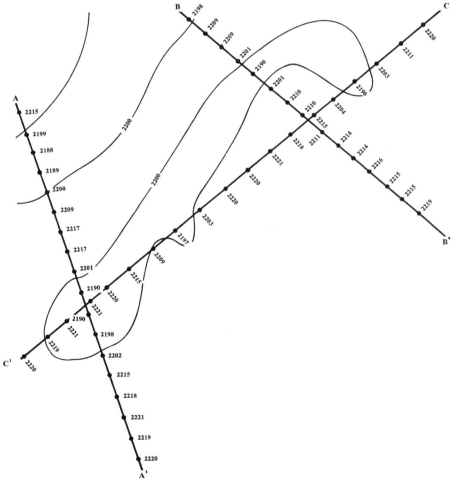

Figure 18. Contouring interpretation B.

Examination of topographic features

So it is sometimes difficult to correlate recorded echograms with given geological features because of this dependency on the reference chart. Fortunately, for positioning and retrieval of position, the peculiarity of a section of echogram need not always be associated with a known *type* of underwater feature. The echograms on the pages that follow illustrate this point. These echograms are considered good; they are clear and suitable for detailed study.

Maintaining a clear echogram trace

It should be noted that, in areas of complex topography, ship's speed must often be adjusted to obtain a clear echogram trace. Slowing the ship by

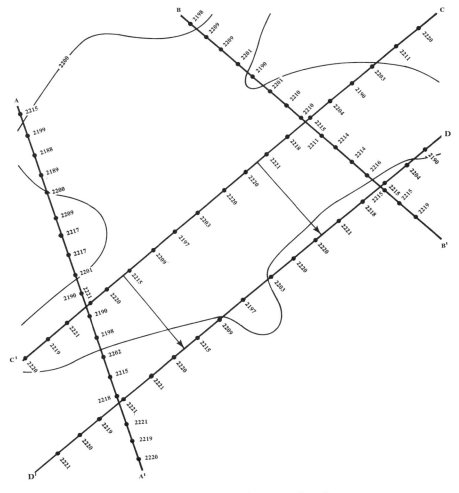

Figure 19. Contouring interpretation C.

several knots will often bring in a good trace. On very steep slopes, slowing the ship will allow some returns from microfeatures at a right angle to the outgoing acoustic signal, even if the angle of the slope itself is greater than half the normal beam width. (At such an angle the returning sound does not normally record.) Controlled outgoing pings will allow the sounding to show in the area of outgoing signal. At a 400- to 800-scale range, for example, depths from 400 to 430 fathoms or so will be obscured by the outgoing signal (or by the deep scattering layer) unless ping frequency is adjusted. This is an example in which the keying of the transmitter and gating of the receiver are used. In passing from deep to shallower water (or the reverse) this is ordinarily no problem, but on surveying in waters of average depth, such adjustment is mandatory unless a change in scale can be shown on the particular echo sounder used.

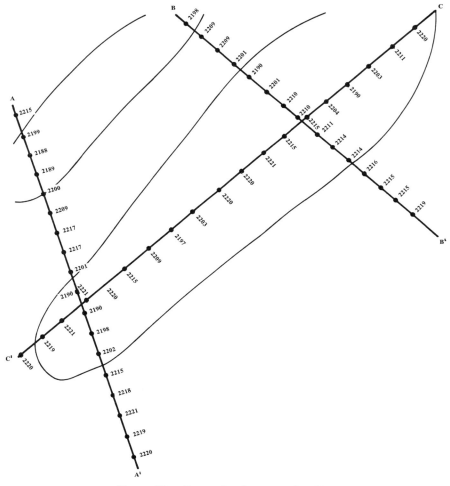

Figure 20. Contouring interpretation D.

Vertical exaggeration

Before examination of the echograms is undertaken, recall that the bottom profile is shown at a vertical exaggeration. As stated previously, the vertical exaggeration of a given section of echogram depends on several factors, including ship speed, scale, and types of equipment used, but in general it is at least 10:1 and in some cases as high as 30:1. One of the techniques by which position recovery is achieved is dependent on profile comparisons, and it would be important to recover position at the same ship speed as that yielding the reference profiles so that the same vertical exaggeration occurs. This is described more fully in chapter five.

Seamounts

Figure 21 shows a well-defined seamount. From the standpoint of positioning this feature may or may not be easy to locate, but once it is, there would appear to be no difficulty in deciding that it is this and only this feature that the ship had found. Isolated seamounts or seaknolls are ideal for such positioning, whereas similar features in groups would be less useful for this purpose. Unless some difference in shape distinguished one from another in this latter case, the remaining identifying factor would be minimum depth. This, in turn, would depend on horizontal distance between the ship and the seamount's top projected to the surface.

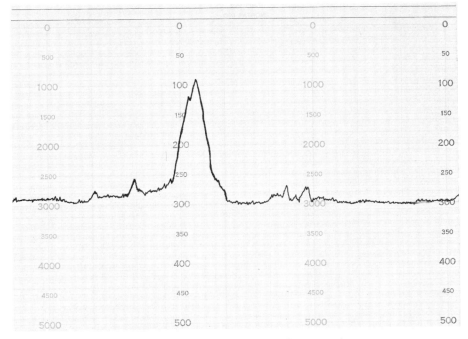

Figure 21. Well-defined isolated seamount.

The fact that such a feature is indicated on the recorder at all means that some part of it lies within the sound cone, but that part need not include the top itself. Figure 22 shows the sound cone as it would appear from the vertical and shows the top of a seamount not being recorded. The resultant echogram would show a seamount (or seaknoll) nonetheless, with minimum depth deeper than that of the reference chart. Here again, caution must be observed, for the reference chart was based on a survey wherein the minimum depth of the feature was noted assuming that the ship passed almost directly overhead. The known width of the sound cone must also be taken into consideration. In the example shown, the ship's recorder would show a peak of about 3,000-fathoms minimum depth.

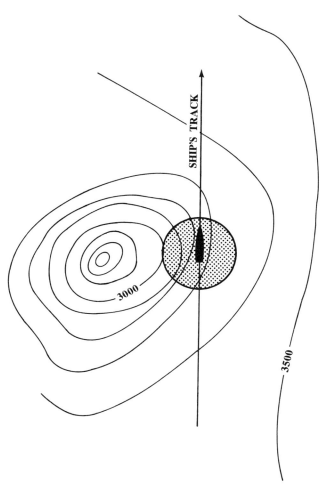

Figure 22. View of sound cone from above. The coverage on the bottom is represented by the concentric circle. It can be seen here that the seamount top is not recorded as the ship passes by.

Escarpments

Figure 23 is a UQN recording which, because of the vertical exaggeration involved, appears as an escarpment-like feature. It is plainly unique in one line of direction and is associated with a flat area. The feature is about 1,500 fathoms high in relief and extends for about 40 nautical miles.

Shallow features

Figure 24 shows a feature rising to within 160 fathoms of the surface. Two multiple returns are seen clearly. Each is incremental; the first multiple return is 320 fathoms, etc. Note the markings *600 feet* and *600 fathoms* at the top of the recording. When the horizontal line crosses the 600-foot marking, the vertical scale is 0 to 600 feet; and when it crosses the 600-fathom marking, the scale is 0 to 600 fathoms (as in this case). A scale of 0 to 6,000 fathoms is indicated when the line crosses both of the markings.

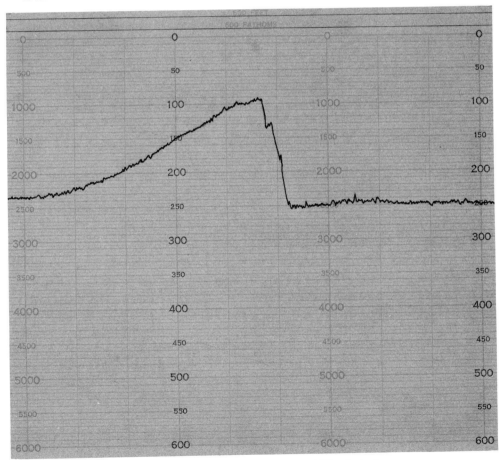

Figure 23. Recording of escarpment-like feature.

46   BATHYMETRIC NAVIGATION AND CHARTING

Abrupt emergence of features

The enlarged recording in figure 25 shows a feature emerging abruptly from a flat area (the track of the ship is from right to left). Spaced sharp

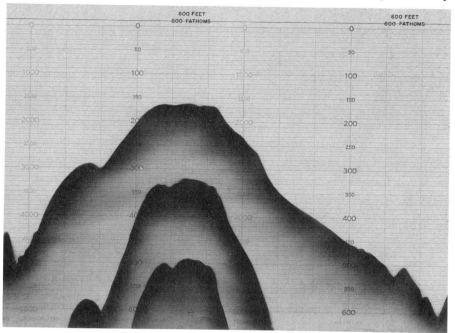

Figure 24. Underwater feature with incremental multiple returns.

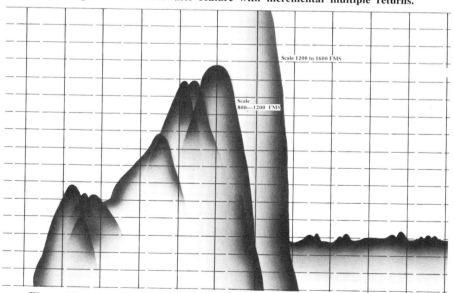

Figure 25. Seamount. A large feature emerges abruptly from a relatively flat terrain.

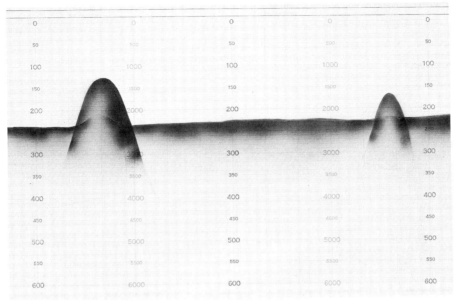

Figure 26. Spaced protuberances. These two spaced features of different minimum depths offer a good choice for locating position based on echogram comparison.

protuberances of five to eight fathoms are seen on the flat area. These are hardly microstructures but are small compared to the seamount. On this recording the horizontal, broken lines are three minutes in duration and are spaced vertically at 20-fathom intervals over a 400-fathom spread. In this case, the echo sounder operator has also indicated five-minute fixes, as seen by the dark vertical lines. This is done by pressing a button on the unit, which energizes the stylus for the length of the recording.

Spaced protuberances

The slight protuberances in figure 26 appear to be off to one side of the ship. Their shape is not of great significance because the effect is hyperbolic. Differently shaped features would thus cause identical hyperbola if the minimum depths were the same.

Submarine canyons

Submarine canyons are found on most continental slopes and are often V-shaped as seen in figure 27. They are relatively steep-sided and show a steadily descending axis or trend. These features are among those most easily recognized on an echo sounder's recorder, especially when the crossing occurs at a right angle to the axis. Since the axis of a canyon descends more or less continuously, particular depth values should occur only at one point along

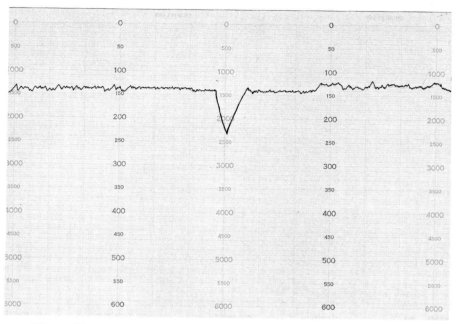

Figure 27. Submarine canyon. This submarine canyon crossing shows a typical V-shaped valley.

that axis or trend. A farther crossing of this feature, in figure 27, some 4,000 yards away would reveal another depth value which, when correlated with the chart, would fix the ship's position. Although more than one crossing of a canyon will show its slope if the crossings are properly spaced, it is not unusual for a single crossing to show a unique depth value, thus establishing position of the ship.

Lows

Two spaced lows are seen in figure 28. The depth of the lows is important, but so are their shapes. These circumstances, plus shape and depths of the tops (seen at a different scale value), are often ideal for correlation. Since there is no indication on the trace, it is assumed the ship had not turned about on a reciprocal course at approximately 0920. If the ship had, this would be the same low recording again.

Plains

Figure 29 illustrates the lack of features sometimes to be found. This flat area is typical of many areas. The bottom here could be fairly shoal or at abyssal depths; if deep, abyssal plains can be found not varying in depth by more than one or two fathoms for many miles. Sometimes they are terminated by sudden changes in slope, and these are significant when well charted.

SIGNPOSTS ON THE OCEAN BOTTOM 49

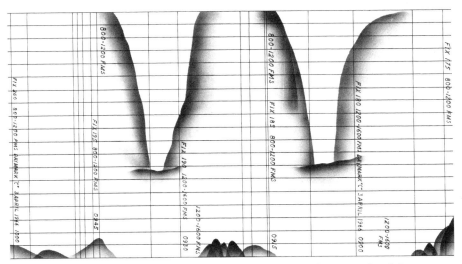

Figure 28. Two spaced lows. Two lows whose shapes and minimum depths, as well as spacing, would indicate position.

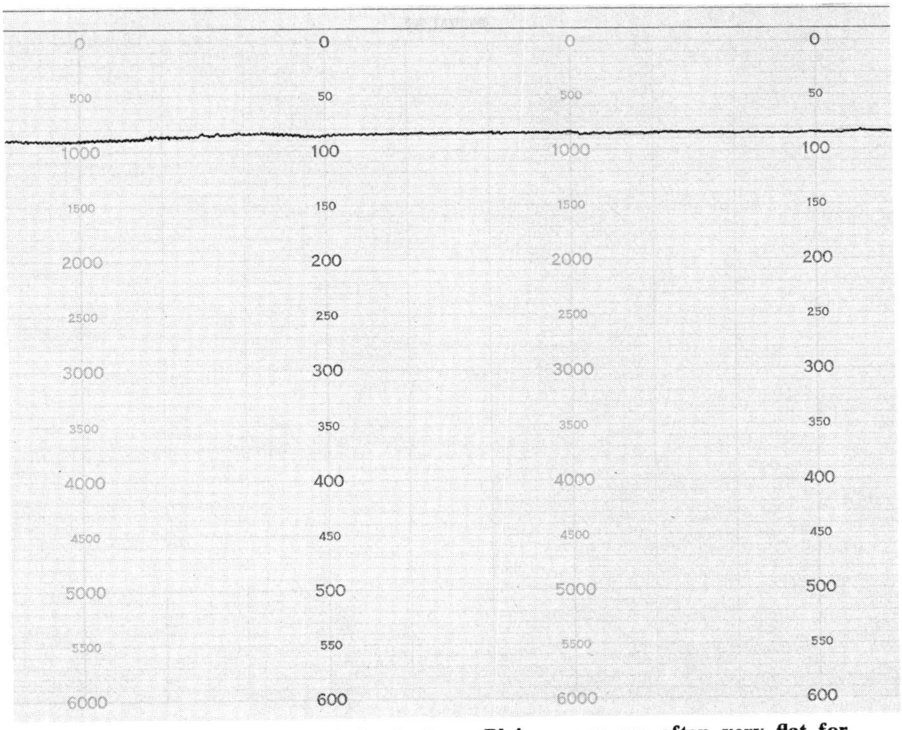

Figure 29. Flat area of the bottom. Plains areas are often very flat for many miles; their termination is sometimes abrupt and indicative of line of position.

Illustration of echograms

The echogram illustrations presented in this book posed difficulties in exact reproduction because of varying quality of originals. In order to bring out the salient points of the original, different techniques were used by the illustrator. Some of the echograms were particularly suited to air-brush application since intersections of the bottom and significant shading changes on the original chart paper could best be shown in this manner. In other cases, the bottom profile alone sufficed to illustrate key points. Experienced oceanographers will recognize that echograms recorded in the original on a large-scale recorder have been presented on a trace which is smaller than the original; in this transfer, scale changes were made to conform with the depth range. While several of the echograms may have been suitable for direct reproduction, it is believed that the combination of air-brush and profile drawings will be more satisfactory for the kinds of detail discussed.

The information in this chapter on geological features and representations of bottom features on echograms is preliminary to an understanding of bathymetric navigation and charting. Familiarity with the echo sounder under all conditions of use and with the nature of the features recorded is essential. A discussion of charts and charting methods will follow. Recognition upon an echogram is but the first step towards correlation, and it is the bathymetric chart that serves as a tool for this purpose. In figure 30 the bottom profile of the ship's track is shown pictorally beneath the track; eventual goals are accurate navigation from one point to another, ability to return unerringly to a particular geographic location, and precise determination of location at a point along the track.

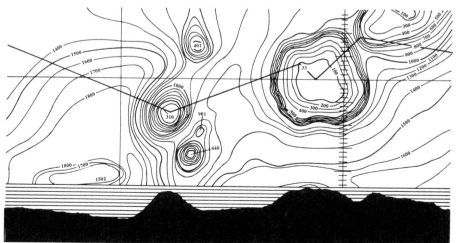

Figure 30. Ship's track on bathymetric chart with associated bottom profile. Bottom features sometimes lend themselves to point-by-point navigation, as shown by the ship's track (solid line), with change in direction determined by the ability to identify the recorded feature. The track's profile is shown below the chart.

CHAPTER IV

Charts and Charting

The word *chart* goes back to the Greek word *chartas,* which meant a leaf of paper. The term *bathymetric* is a modern application to a type of chart, the derivation of which is almost self-evident; *bathy* is also from the Greek, *bathos*, meaning depth. The word *map* is from the Latin word for napkins, *mappa*, derived from Carthage where it was used to mean signal cloth. Maps and charts represent, upon a flat surface of paper, a portion of the earth's features and may also bear other notations for purely functional reasons. Maps may show features that are geographic, political, economic, demographic, biogeographic, meteorologic, geologic, and all the features that a culture may delineate. Charts usually show only those items considered significant for the expressed purpose of the chart. Some make the distinction between a map and a chart that a chart is merely a specialized form of map.

Still another point must be considered when speaking of the term *map* either as a noun or a verb: one can *map* oceanographic properties other than bathymetry. Geophysical characteristics such as gravity and magnetics are also *mapped*.

Nautical charts, the larger grouping of which bathymetric charts are a part, contain whatever information is necessary for safe and efficient navigation upon oceans and inland waterways. Bathymetric charts refer to topographic charts of areas in the deep ocean and of the ocean floor. In the past fifty years the development of new techniques of accurate navigational control and of better methods of sounding has gradually moved surveys from bays, shelves, and near-shore areas to the deep sea.

The purpose of this chapter is to acquaint the navigator with survey and charting principles to make the bathymetric chart a more useful tool for general navigational positioning. The finished chart is the culmination of much effort, and, to the extent that these efforts are successful, the bathymetric chart will have utility and reliability for navigation.

Compromise is sometimes necessary in chart production because some factors preclude presentation of all of the data that can be collected in a given area. The information shown must be so presented that it can be understood with ease and with clarity. Incorporating upon a chart more than a necessary amount of information (depending upon its nature and purpose) might increase accuracy at the expense of utility. The decision on what should be and what can be portrayed is one often faced by the chart producer. Other than the nature of the information and its attainability to the accuracies required, other factors contributing to the decision as to what is shown on the printed chart are scale, utility, and topography, in varying degrees of importance.

The first formal Hydrographic Office was established by Spain in 1508 as the "Casa de la Contractacion de las Indias." This was primarily a commercial enterprise designed to regulate overseas trade. A separate department of the Casa was organized to chart the New World areas and the routes to and from the New World. Despite the undisputed scientific value of the findings, these efforts were not entirely disinterested. Spanish commercial voyages of the times required more navigational information than had previously been available. Much of this information was kept secret because a safe and rapid sea route was frequently the road to great wealth.

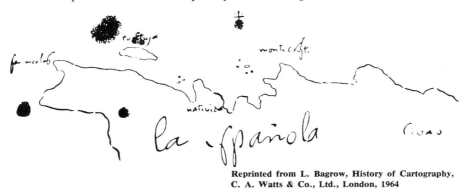

Reprinted from L. Bagrow, History of Cartography, C. A. Watts & Co., Ltd., London, 1964

Figure 31. Manuscript sketch chart of northwest coast of Hispaniola by Christopher Columbus in December 1492.

There are many old charts extant, one of the more intriguing, perhaps, being a sketch made by Columbus himself of the northwest coast of Hispaniola (figure 31), the one surviving chart of several he was supposed to have drawn. Only two other charts remain from his voyages, drawn by men who sailed with him.

Chart portrayal and preparation for surveying

On bathymetric charts, relief is generally portrayed by contours, which are lines connecting points of equal displacement (depths) measured from sea level as the accepted reference datum. The chart datum is the permanently established surface from which soundings or tide heights are measured (usually low water). Other schemes which have been used to show relief are hachure (shading) marks, tinted layers, line drawings, and physiographic diagrams. Some of these portray relief quite forcefully. Most modern charting methods use contours that allow for determination of depth anywhere on the chart, even between the contours. The regular interval used allows depths to be determined, contours to show shape, and slopes to be measured.

Production of a bathymetric chart begins with the survey and even before then a great deal of planning goes on to ascertain how the survey will be organized. There is no single correct way to survey an ocean area to produce a chart. Although certain procedures have been established, they are considered to be guidelines that very much depend on given circumstances. Some factors that will determine specific procedures to be used are the kind of navigational control available, reliability and range of navigation, the average depth of the water, the kind of topography expected to be found, the scale of the chart to be produced, the equipment on hand to do the job, and the number and kind of sounding platforms (buoys, helicopters, ships, soundboats, etc.).

Classification of bathymetric charts

Bathymetric charts can be classified as controlled or uncontrolled. The *controlled chart* is produced as a result of a survey of specific areas, usually an intended series of charts. Charts that are produced from surveys are usually large-scale, are based on reliable control, and have had the area covered systematically. The contour interval is probably large, ranging from five fathoms to 50, commensurate with scale. A bathymetric chart at a scale of 1:25,000 may show a 20-fathom contour interval, and another at 1:50,000, a 50-fathom interval. The interval is dictated in part by the nature of the topography, but in general the closer the interval, the larger the scale. In areas of complex terrain with numerous highs and lows, a smaller contour interval can suffice, although in such a situation the ability to measure larger intervals from echograms is probably greater. In flat terrain the closest interval possible, i.e., five fathoms as opposed to 25, is desirable.

Uncontrolled charts are based on random data collected from ships that have traversed an area on their normal steaming route. Because of poor navigational control and the scale of plotting sheet on which the track is maintained, the reliability of the data may be poor. These charts are almost always at a smaller scale and a greater contour interval. Such a chart at a scale of 1:650,000, for example, would normally be found with a 100-fathom contour interval. The chief characteristic of the class of data that goes to produce such a chart is the difficulty in reconciling the conflicting depth and positional information from many sources. Random track charts of ships, whether naval or merchant marine, show frequent inconsistency in depth and position. This is not to say that larger scale charts are always produced from survey data alone or that small-scale charts do not have survey data incorporated into them. Certainly this is not meant to certify consistent reliability of survey-type charts as opposed to those produced by random data. Table four summarizes the general characteristics of the two types of chart products. Table five relates different scales to each other in terms of linear measurement.

TABLE 4.

General characteristics of controlled and uncontrolled chart products.

	Scale	Contour Interval	Surveyed	Reliability	General Availability	Suitability for Bathymetric Navigation
Controlled	1:10,000–1:100,000	5–50fms	Yes	Generally good	No	Good-excellent
Uncontrolled	1:1,000,000 1:1,500,000	100fms	No	Generally poor-fair	Yes	Fair

TABLE 5.

Representative linear measurements at different scales.*

Chart Scale	One Inch in Nautical Miles	One Nautical Mile in Inches
1:5,000	One inch equals 0.069 NM	One NM equals 14.58 inch
1:10,000	0.137	7.29
1:20,000	0.274	3.65
1:25,000	0.343	2.92
1:50,000	0.686	1.46
1:100,000	1.371	0.73
1:250,000	3.429	0.29
1:500,000	6.857	0.15
1:750,000	10.286	0.10
1:1,000,000	13.715	0.07

* At identical latitudes depending on cartographic projection chosen

With regard to present availability of the two products, small-scale world coverage is available in several chart series. Controlled charts exist for relatively few areas of the world. The situation is one that is being improved steadily in that more surveys are being undertaken; these surveys in time will result in greater numbers of controlled, larger scale bathymetric charts.

Survey techniques and principles

Characteristics of a given survey depend upon several key factors. The number of passes, or sounding lines run by the ship, depends on average water depths and the echo sounder used. Average depth is in turn affected by the kind of topography found in the area. A transducer utilizing a narrow sound beam necessitates a greater number of sounding lines for coverage of the area. Much interdependence exists. The scale of the chart may be the determining factor on line spacing if other factors do not preclude such consideration. Although complete bottom coverage of an area is normally desired, this may be dispensed with on occasion. Again, the decision to survey an area without complete coverage would be dictated by average depth, transducer characteristics, and topography.

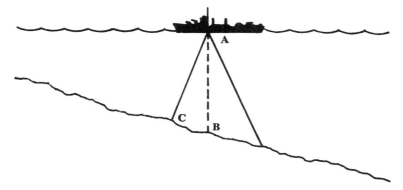

Figure 32. Correcting for slope effect to find absolute bottom depths. Point C is being recorded as the depth in this illustration although the actual depth under the ship is Point B. On a regular slope of this kind, the true depths can be found through correction procedures.

Projection of cost or time for given surveys may be determined by linear miles to be surveyed. Variables that must be considered for this determination include number of ships available, track spacing (in turn dependent upon depth and transducer characteristics), and average speed of the ships. However, it is not unusual to find that control reliability and weather conditions in the area will require modification to specifications developed beforehand.

The specific *techniques* of survey operations and chart production are less important than understanding their governing *principles*. Manuals of hydrographic and bathymetric survey techniques are published by the Naval Oceanographic Office of the U. S. Navy and by the Coast and Geodetic Survey of the Environmental Science Service Administration (ESSA).

Correction for slope and sound velocity differences

The vast majority of existing data has been accumulated by using nondirectional transducers, and the depths shown at given points in the echogram are consequently subject to slope effects and side echoes. In complex topography it is virtually impossible to correct for slope effects, but on simple slopes correction is possible. In figure 32, the distance AC is being recorded over actual depth AB, since at any moment C is the point on the bottom nearest the transducer. Correction methods exist, both mathematical and graphic, which yield absolute bottom depths in such a case. Differences in sound velocity may also be corrected since the observed velocity sometimes differs from that with which the ship's equipment is calibrated. In surveying for purposes of installing gear on the bottom and for determining absolute depths of certain shoals, the corrected depth is vital. For navigational purposes, however, corrected depth is not as important because echo sounders in all U. S. Navy ships are calibrated for the same sound velocity of 4,800 feet per second and because the errors, if any, would be repeatable and could be discounted. It does not matter if bottom topography is portrayed uncorrected, provided the ships that follow also do not correct for differences in sound velocity and depth and have similarly calibrated equipment. The velocity of sound will also show a variation with time in certain areas, but this alone is unlikely to cause major changes in relief portrayal.

Primary and supplemental lines of survey sounding

Precise direction of the survey runs made by the ship is often related to the contour trend in the area. The few formalized survey instructions that do exist state that primary lines of chart development should be run at right angles to the general contour trend and that supplemental lines, at perhaps ten times the usual track-spacing (one-tenth the number) should be run in the direction of the contour trend. It is the author's belief that primary lines should be run parallel to the contour trend and supplemental lines should be at right angles. By surveying in a direction up and down slopes, which is the situation that occurs running at right angles to the trend, (except in very complex topography), a more detailed trace usually results. This happens because the leading or the trailing edge of the sound cone (depending on direction of the ship up or down slope) is recording the depth values. Values constantly shoaler than those directly under the ship are being recorded. Although simple slope correction procedures may be applied, the data recorded are oversimplified. Side echoes, if present, would not record as well—the good, clear trace that usually results in an indication of this. By steaming parallel to the contour trend, more data are recorded; although conflicting data, nonetheless more data. (In theory the shoaler depth is still being recorded, if the slope is great enough, even by steaming parallel to the contours). These data may cause more difficulties in interpretation, and some cross-check lines may be incompatible. The choice, however, is whether one wishes to record less data which offer little possibility of conflict in interpretation and which are easier, therefore to contour (only a limited choice of patterns of contours would be available), or whether one has the opportunity to record all of the possible data, conflicting and difficult to contour readily, but capable of maximum interpretation. The difference, in any case, is usually important only in a certain kind of area, such as where island slopes dip steeply but are interspersed with low hills, knolls, or canyons. This configuration occurs frequently on ridges affected by past volcanic activity.

Navigational control

The navigational control of the survey is paramount; its accuracy determines survey ship position at all times so the depths may be annotated in proper geographic location. The range of the shore-based systems affects the distance from the control stations at which the survey ship can receive its navigational input accurately. Many electronic control systems are available to choose from, but given needs of a particular survey dictate use of specific ones. Control systems such as Lorac, loran, Decca, Raydist, shoran, and Hiran are well known. The earlier circular systems associated with single stations have been superseded by hyperbolic phase-comparison systems. A typical pattern of hyperbolic lines of position is shown in figure 33. Loran-C is a hyperbolic system that allows optimum balance of accuracy and range. Other control systems now in the developmental stage will offer good range, and satellite navigation is available. One characteristic of a given control system that may determine its use on a particular survey is its ability to support multiship operations; some can and some cannot. As an aid for survey work, electronic control systems are indispensable. The greatest need for support of navigation using bathymetric means is for more data resulting from adequately controlled surveys.

Surveying

The actual mechanics of the survey are straightforward. This involves steaming along prescribed tracks laid out beforehand, ascertaining the reliability of the echo-sounding trace, and measuring the echogram at a prescribed time or depth interval. Usually the scale and nature of the topography indicate whether it would be best to measure the depth as a function of time or at every change in depth, say of five fathoms. Careful records of the ship's track are maintained during the survey, and sheets of depth values properly positioned are eventually prepared for the contouring phase.

Contouring

Contouring is one of the more crucial phases of bathymetric chart production, and upon the validity of the final contours rests the combined worth of all the previous effort. It is a skill which not everyone can learn. Success in contouring depends heavily on knowledge of landform (in the generic sense), and a background in geology—preferably submarine geology—is indispensable.

Scale

Before the nature of contouring is discussed, it is necessary to examine some fundamentals concerning scale. Coverage of common large-scale charts in square nautical miles is shown in table six.

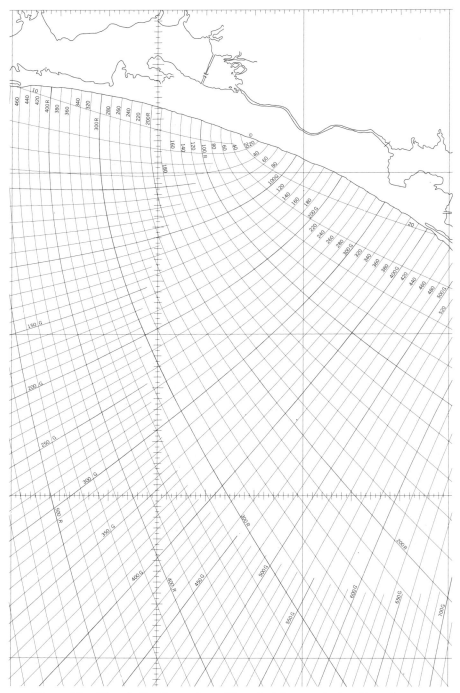

Figure 33. Hyperbolic lines of position depict a control pattern from electronic positioning stations.

TABLE 6.

Area coverage at various scales.

Scale	Coverage (Square nautical miles)
1:10,000	24
1:20,000	108
1:25,000	165
1:50,000	682
1:100,000	2,340

It is scale and topography that contribute to the choice of the contour interval. To show the importance of scale, a 100-fathom contour interval on a large-scale chart of flat topography could very well show no contour lines.

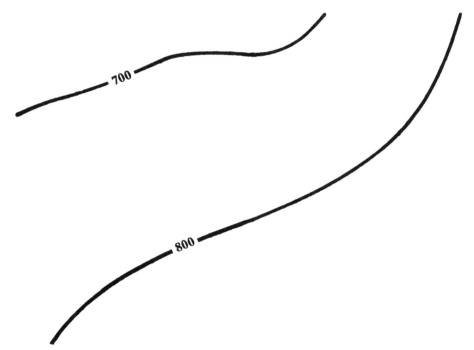

Figure 34. A 100-fathom contour interval. This interval on a small-scale chart could provide very little useful bathymetric information.

Conversely, a five-fathom interval on a medium- or small-scale in complex topography would be a maze of indistinguishable lines. Since it is by contours that bathymetry is shown, it is by the closest contour interval that we wish to position a ship bathymetrically. Figure 34 at a 100-fathom contour interval would offer little data for positioning purposes. The same area at the 20-fathom interval gives more of the detail necessary (figure 35). No useful detail can be readily seen in figure 36, where the inset (figure 37) is shown enlarged at the 10-fathom contour interval.

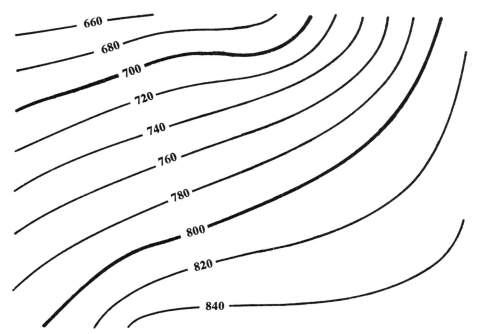

Figure 35. A 20-fathom contour interval. This closer contour interval displays more information about the same area.

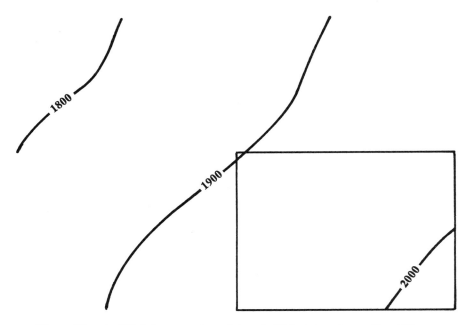

Figure 36. A 100-fathom contour interval. This interval provides little data for positioning purposes in this kind of topography.

Contouring and terrain interpretation

The arrangement of contours in the following illustrations appears faulty. A doubtful pattern of 20-fathom contours has been constructed in figure 38, for example, by the appearance of the 420-fathom contour. Perhaps this specific contour is drawn correctly, but if so the remaining contours are drawn incorrectly. One kind of topography would undoubtedly prevail, especially in the limited area described by the contour interval.

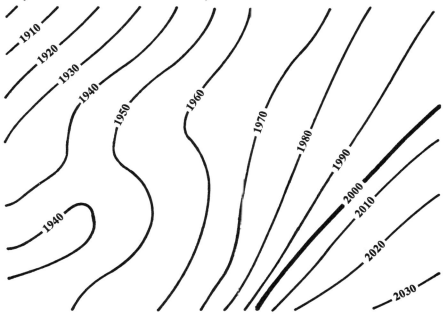

Figure 37. A 10-fathom contour interval. When the inset from Figure 36 is contoured at a closer interval, a more complete picture of the bottom emerges.

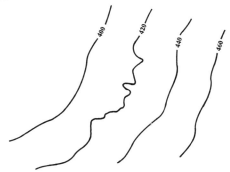

Figure 38. Doubtful pattern in a 20-fathom contour interval. The 420-fathom curve is here apparently misdrawn. If the 420-fathom curve were correctly drawn, the remaining contours would be wrong. There is no consistency in the pattern as presented.

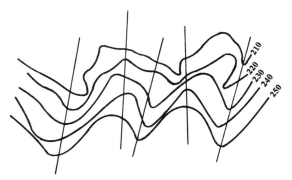

Figure 39. Loss of geological integrity. The axes of ridges and lows (shown by the straight lines) are not carried through in this illustration.

Figure 39 is a contour arrangement of slight linear ridges and lows. Note that these have not been carried through, however, and that the contours are fragmented. One or two of the contours may be correctly positioned. If the axes of the highs and lows as those shown by the lines were considered in the interpretation, contours about these axes would offer a more realistic picture of the bottom at this point. In aligning the axes and contours it is entirely possible that none of the resultant contours would be as accurately positioned as one or two may have been in the illustration. Nevertheless, a more complete and realistic portrayal of the geological relationship would probably result. Figure 40 appears readily acceptable; the double canyon-like lows meld into one low topographically, and the convex shape of the 240-fathom line may indicate an accumulation of debris from slumping.

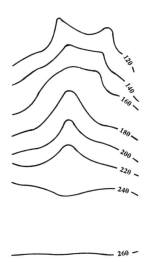

Figure 40. Proper contouring. This pattern is readily acceptable, showing two lows melding into one and a possible accumulation of debris near the bottom flat.

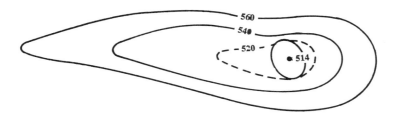

Figure 41. Seaknoll based on given contour and more probable shape. The dashed 520-fathom curve is probably correct in view of the shapes of the other contours and the position of the 514-fathom minimum depths.

The knoll seen in figure 41 would present a more realistic picture if the 520-fathom curve were seen as depicted by the dashed line. Too abrupt a change in direction would result if the solid 520-fathom curve were used. In looking at this from a side viewpoint as illustrated in figure 42, the difference d represents the alternative positions horizontally of the two schemes of contours. Profile A is more likely than Profile B, for it carries through sequentially with the slope angles on either side. The high point (minimum depth) should not be equidistant from the two contour intersections with the slopes because of the different angles of those slopes.

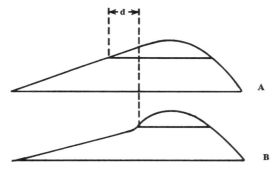

Figure 42. Side view of Figure 41. Either of these shapes is plausible as a geological feature, but, in the absence of information indicating otherwise, A is more likely the shape to be achieved by contouring.

Finally, figures 43 and 44 indicate by dashed contours what the more desirable representation of the bottom should be. In both cases the existence of the tops is inferred by the dashed contours; actual examination of the echograms would be required to validate the existence of the tops.

The first attempt at contouring an area can be based exclusively on the depth data as positioned on the survey tracks, but it cannot be left to this alone. At times, use of depth values alone often allow for alternative shapes to emerge, each perfectly consistent with the data. Interpolation is not so much the problem as is interpretation. General highs and lows should be carried through up or down slopes unless there are firm indications to the contrary. A decision that a given contour is incorrect, though based upon the soundings shown, may be quite in keeping with interpretation by trend; any one of a number of factors may have occurred to make this decision.

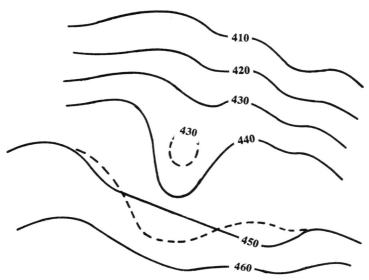

Figure 43. Incorrect contour portrayal. The knoll in this illustration has not been carried through. It is possible, but doubtful, that the feature terminates as shown by the solid contour below. The dashed contour is probably more correct.

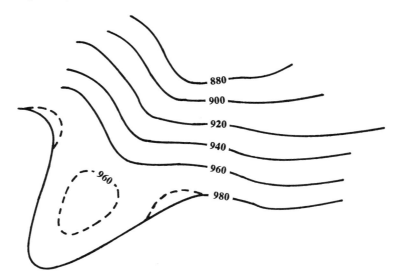

Figure 44. Differences in interpretation. Either the dashed or solid line is correct. Individual interpretation might dictate that the dashed contours are more desirable.

Utter disregard of the plotted depth values on the adjusted survey ship's track is obviously not desired. But within the range of errors normally found, interpretation is more than desired, it is mandatory. This is true even to a greater degree in the case of uncontrolled surveys.

How much data are needed to render contouring accurately varies with terrain. Over a certain kind of topography, more than a given amount of data (assuming adequate position control) will not yield corresponding increases in accuracy. Over very complex terrain it is doubtful whether instrumentation limitations allow for accurate portrayals even if an infinite number of sounding lines were surveyed. Here it is probable that more than a certain optimum bank of data will cause difficulties. This would be due to the profusion of conflicting depths caused by side echoes, a situation which cannot be completely overcome even with the best of positional control. The problem sometimes resolves itself, for the control often cannot allow for lines to be run too close together.

Random data interpretation

Even when gently rolling terrain is contoured from random data, the contouring stage sometimes presents difficulties. Figures 45 to 47 show different examples of contouring a given set of depth data. Whereas there might be no hesitancy in granting to figure 45 the more pleasing and apparently authoritative set of contours, one or another of the three attempts could be as accurate as the other *at given points*. The overall shape and kind of bottom would seem to favor figure 45; and, indeed, this is the more professional rendition. But were figures 46 or 47 published as the final version, no great difficulties would be encountered in recovery of general position in this case—to the extent there were no other precluding factors present. Indeed, closer examination of figure 46 will reveal that this example has the depth information plotted only at 20-fathom contour *crossings*, with highs and lows as appropriate. The cartographer measuring the echogram was presumably aware that contouring would be achieved at a 20-fathom interval. Given the contour crossings alone, it is a relatively simple matter to connect the values in approximate shapes. In very steep areas of topography, this is quite agreeable. Yet figures 45 and 47 that show the depth values measured as a function of *time* present the more meaningful—albeit more difficult— data base from which to work. While it is true, as previously stated, that in areas of complex terrain more than a given amount of data poses difficulties almost in direct ratio to the amount of additional data beyond an optimum figure; in other cases the professional cartographer wishes to avail himself of all the data that he possibly can. Conflicting information may in itself be of significance to an experienced cartographer, as might the lack of conflicting data where one would normally expect to find it.

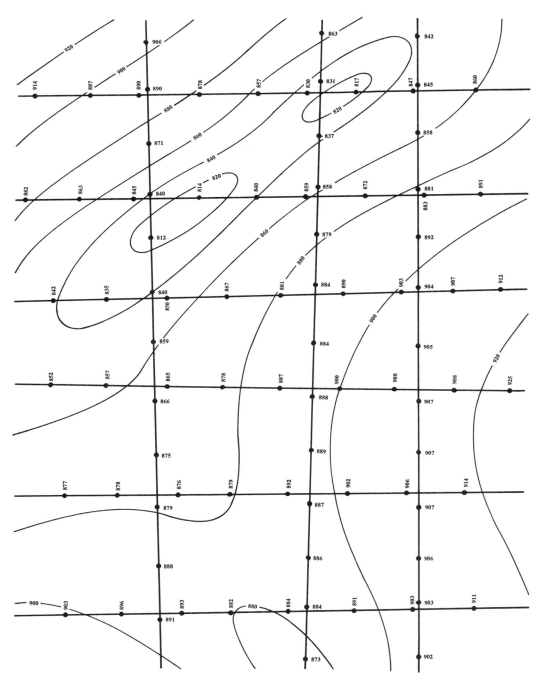

Figure 45. Contouring interpretation A of an area with a given set of depth data.

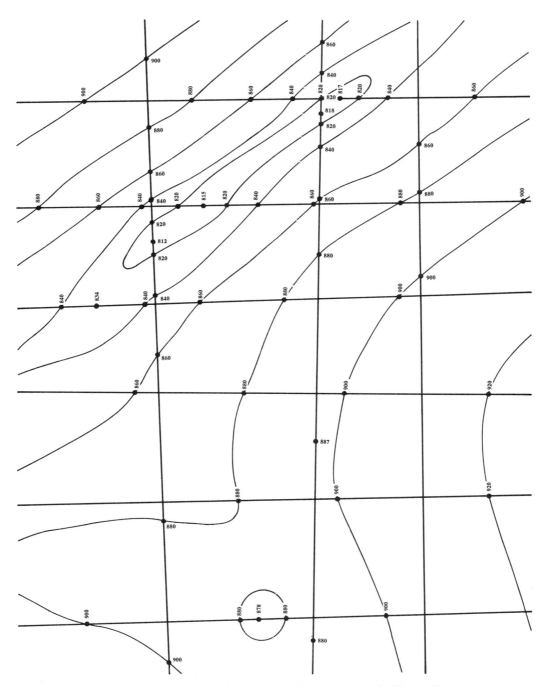

Figure 46. Contouring interpretation B of same area in Figure 45.

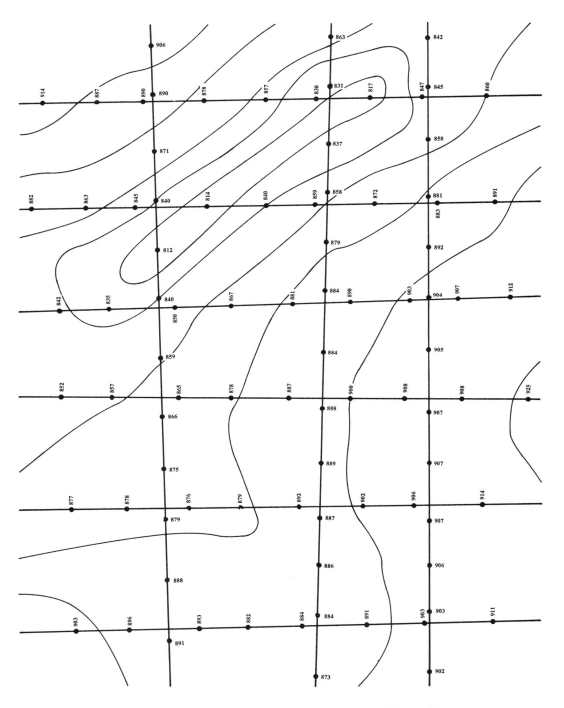

Figure 47. Contouring interpretation C of same area in Figure 45.

Survey data interpretation

In an attempt to show different interpretations of identical data by different individuals, the experiment depicted in figures 48 to 51 was devised. The depth data do exist in that they were annotated from a published set of contours. The portions of the data subject to most interpretation are in the vicinity of the low in the upper central portion and the slight knoll in the lower right portion of the chart.

Four well-qualified individuals were chosen to contour the area at a 20-fathom interval. The results are shown in figures 48 to 51. It is important not to draw too many conclusions, as one might be tempted to do by correlation of the efforts with individual background and experience.

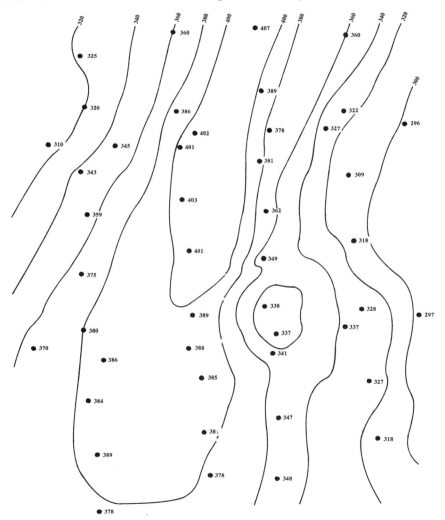

Figure 48. Contour interpretation experiment—First man.

70 BATHYMETRIC NAVIGATION AND CHARTING

More complex experiments could be constructed that would compare contouring as a function of time, use incorrectly plotted ship's tracks, or compare a published chart or other reference. Another could be devised to gauge qualitatively the minimum amount of data that must exist for all individuals to come up with similar contours. Yet another might compare computer-drawn contours with manual sets under differing circumstances.

As to this particular experiment in figures 48 to 51, only a limited number of conclusions can be drawn. Examination reveals, first, that no major difficulties existed in most of the area; it was more a question of connecting points of equal depth, although positions of the contours vary due to interpolation.

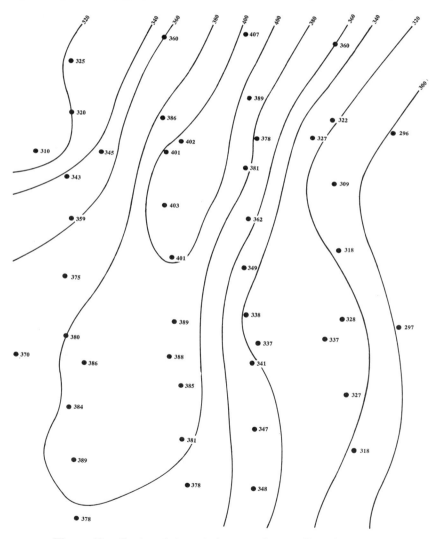

Figure 49. Contour interpretation experiment—Second man.

Second, differences are seen in the interpretation of the center trough-like feature and in the case of the small knoll in the lower right portion of the chart. One valid general conclusion can be drawn from the experiment, one which substantiates what the author has attempted to point out, namely that different individuals will interpret identical data variously if the data are of a nature that is amenable to such differences. To paraphrase Murphy's Law "if something can go wrong it will," it can be stated that depth data lending themselves to differing interpretations will be interpreted differently.

There is a way of comparing the different attempts to some reference set which is a criterion, and this would be the published contours. But here too,

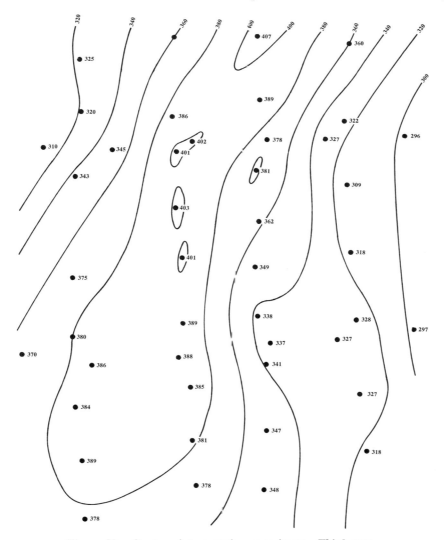

Figure 50. Contour interpretation experiment—Third man.

the published contours must be assumed to be correct. Presumably the published contours would be the best interpretation it had been possible to make given that data. Best interpretation or not, it may or may not represent the actual bottom.

Modified contour portrayals

The existence of some contours merely indicates a prognostication of shape whose correctness depends on how well the interpretive process was accomplished. The need for this process has been established as necessary becaue of the nature of sounding data. Can some sounding data be profitably ignored to improve utilization of the chart product? They can, if the final product is not limited to topographic portrayal by means of contours alone.

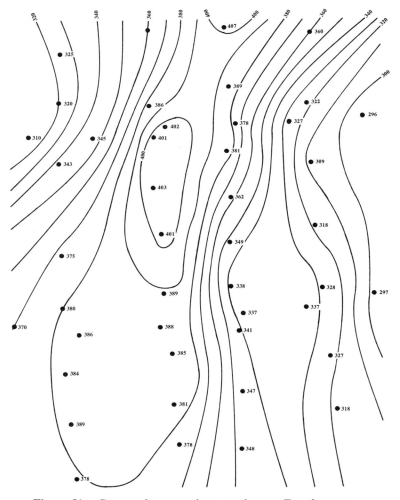

Figure 51. Contour interpretation experiment—Fourth man.

Such a product might display selected contours in prescribed interval or depth range and could also depict ocean floor provinces as a function of rate of change of depth per given distance. This would preclude the necessity of delineating specific shapes of features by minute variations in the contours where no substantial basis for such detail is present. A variation of this technique could show echo-sounding profiles across the area of the chart, obtained from original survey records rather than from previously contoured information. The confidence factor of the navigator would be greater if he knew that only those contours considered by the cartographer to be correct were on the chart.

Flexibility in field procedures

Several years ago the author was responsible for surveying an area for the purpose of placing a large piece of equipment on the bottom which was at 1,700 to 2,000 fathoms. Based upon the topography as found on existing charts, overall depths, and the control system available (Decca), several chart scales were chosen for development. The final product was to be at 1:10,000-scale, with five-fathom contour interval. The rather close interval was chosen because, although the slopes were steep, they were interspersed with complex knolls and depressions. During the survey phase it was found that reliability of control allowed only 500-yard spacing with any repeatability. After this primary development was completed, the ship attempted to split the existing lines, that is, steam between the first group of runs to yield a closer pattern. The topography was found to be much more complex than anticipated, and it was decided to develop cross-check lines at the minimum spacing possible. Control was generally reliable, i.e., it was repeatable. Despite some of the conditions encountered there was no reason to believe that contouring of the data would present more than normal difficulties.

It was found, however, that the data plotted were almost impossible to analyze. Conflicting data were present in quantity, and depths from different tracks differed at intersections by as much as 50 fathoms. This was extremely large, for the overall relief covered by the 1:10,000 sheet was only about 350 fathoms. A series of contours was finally agreed upon after further supplemental data were obtained by steaming diagonally across the area, and after discarding several lines which were suspect. The precise picture of the bottom was fortunately of secondary concern since what was desired was a flat area of given slope for the implantment. With the equipment in use, the control, and the complex topography, this set of contours was the best possible.

In retrospect it was determined that the echo-sounding traces were too dark; the set had been adjusted so that at times side echoes recorded faintly or not at all. This alone could have accounted for the incompatability of depth values measured from the trace. Presently, a better set of contours could be obtained for many reasons, not the least of which is the current availability of more directional echo sounders. This example is offered to indicate not only the difficulties inherent in obtaining good bathymetric charts even at very large scale, but also the importance of both careful planning and flexibility during the survey.

Accuracy and choice of charts

To avoid the assumption that all or even most bathymetric charts are inaccurate because of dependency on individual interpretation, it should be pointed out that the contours on printed charts are the best obtainable. They are as accurate as possible, considering their nature and the circumstances under which collected. The data must be approached with realism, with understanding, and with due consideration to instrument limitations, the opaqueness of the oceans, and the survey techniques in use. In the United States, accurate and reliable large-scale charts exist covering many coastline areas. Coverage of continental shelf areas also exists in the sense that everywhere they are represented on some chart series, but such charts are frequently smaller scale and based on dated or erroneous information. Recently, the U. S. Coast and Geodetic Survey has begun publishing excellent bathymetric charts covering the continental shelf off certain coastlines in Alaska and southern California.

If a choice exists, the navigator should avail himself of the largest scale chart available for bathymetric navigation and positioning. Of course, it would be futile to use or to expect to obtain too large a scale of charts when steaming for long distances. A 100-fathom interval chart at medium scale will indicate general highs and lows, specific ridges, some specific seamounts and table mounts, and most canyons. It is superfluous to call to attention the importance of the supplementary information found on most charts regardless of scale, such as date of issue, diagrams indicating degrees of reliability over given sections, or indications of what sections of the chart have been constructed as a result of surveys. The degree of reliance placed upon given charts or portion of charts by the navigator is no less of an art than is navigation itself. With experience, however, comes the skill that will allow the chart to become a much more useful tool.

CHAPTER V

Determination and Recovery of Position

It is important to reiterate the distinction between bathymetric positional requirements for general navigation and for accurate locations. This chapter will discuss techniques which allow accurate locations to be achieved. Needs for greater accuracies in connection with navigation are marginal; the needs are rather for extension of systems on hand to cover more areas of the world and for better degrees of reliability. Certainly it can be expected that the major difficulties in providing dependable long-range navigation will soon be solved. However, very little ship time is lost even now because of navigational uncertainty; on a quantitative basis more delay is caused by weather conditions such as fog or heavy seas.

Gravity observations

The requirements for very accurate position over and above that needed for navigation has been discussed in the Introduction, but one further example may be cited. Cetain measurements are utterly dependent upon ship's position at given times, and here should be mentioned the effect of gravity.

Isaac Newton's universal law of gravitation was based upon observation within the solar system, "Every particle of matter attracts every other particle with a force that varies directly as the product of their masses and inversely as the square of the distance between them." A pendulum clock or a falling rock has a lower acceleration at the equator than at 30° Latitude. By moving from the equator (0° Latitude) to the North Pole (90° Latitude), the speed of the earth's rotation (centrifugal force) decreases, but acceleration due to gravity increases. A pendulum mechanism called a *gravimeter* measures the variation in the acceleration of gravity over the earth's surface; the gravimeter at sea is actually measuring the differences or local variations in the mass distribution in the oceanic crust and the upper mantle. This instrument may be lowered to the ocean bottom in shallow water, or, because of improvements in techniques since the late 1950s, a ship or a submarine may be the platform for the gravimeter. More recent developments involve research into airborne gravimeters and ship-towed gravimeters.

Underway gravity observations at sea are vital to numerous scientific studies and also form an important input to charting world gravity networks. To collect gravity data at sea properly, positional methods are required that are more accurate than normal. Even close to shore, the accuracies needed are greater than nearshore visual bearings and radar-ranging techniques allow. In the absence of more accurate positional methods, uncertainties greater than mean observational errors of the gravity system are usually unavoidable. Much effort is spent adjusting positional data in analysis of gravity data, and the problem is compounded very far from shore.

Positioning and recovery techniques

Through the use of bathymetric information, accurate position can be determined and recovered in several ways. There are those techniques which can yield both lines of position and fixes and which depend primarily on recognition of individual features on the bottom. Examples are crossing of a canyon, location over a specific seamount, and passing over a deep trench. Another category, which includes the method of correlation of observed track profiles with sets of referenced profiles, is sophisticated, is dependent on specified procedures, and yields positions generally of high accuracies. Both categories are described in this chapter. In chapter six are discussed computer-oriented techniques which could fall into a separate category, although because of the strong dependency on correlation they could be considered an extension of the second category mentioned here. Computer-oriented techniques utilize storage and retrieval devices, matrix accumulations, and adjustment of infinite sets of data to yield best-fit information of high accuracy.

Still another group of methods that can yield position could, for want of a better term, be called substantive. They include the many environmental observations sometimes made automatically by navigators. Over flat abyssal plains, for example, the bottom varies in depth no more than a fathom or two for many miles. Yet the boundary of the plains area is often abrupt and quite noticeable on the echogram. This is indicative of line of position. Another example is that of noting a well-defined contour on the chart and on the echogram. Another is crossing a series of tablemounts, (or guyots). Still another is the crossing of a large trench, (such as the Puerto Rico Trench), or a major ridge. These features will not yield position accuracies of a very high order but can be seen by anyone who follows echo sounder data closely. Table seven lists the categories of techniques and the specific ones in each together with their characteristics.

Recognition techniques

The category of techniques that embrace recognition as its primary requirement would be associated with suitable features which are easily identi-

TABLE 7.

Categories of positioning and recovery techniques

	Repeatability [1]	Time to Ascertain	Scale of Chart [2]	Type Chart [3]	Reliability [4]
I RECOGNITION					
Canyons, trenches, seamounts, etc.	±100 yards	1 minute-plus	Large to small	Any	Good
II PROCEDURAL					
Profile matching	±25 yards	1 to 2 minutes	Large	Survey	Excellent
Contour advancement	±100 yards	1 to 2 minutes	Large	Survey	Excellent
Line of soundings	±40 yards	1 to 2 minutes	Large to medium	Any	Good
Side echo	±30 yards	5 to 20 minutes	Large to medium	Any	Good
III COMPUTER-ORIENTED					
Matrix accumulation	Undetermined	Instant	Medium	Survey	Excellent
Other	Undetermined	Instant	Large to medium	Survey	Excellent
IV SUBSTANTIVE					
Plains, tablemounts, large ridges and trenches, etc.	±125 yards	1 to 10 minutes	Medium to small	Any	Fair to good

[1] As opposed to accuracy, this is dependent upon the reference chart in use.
[2] Large: 1:10,000 to 1:50,000; Medium: 1:50,000 to 1:300,000; Small: 1:300,000 and smaller.
[3] From a well-controlled survey or poorly controlled and/or random tracks.
[4] Assuming 75 percent probability of attaining repeatability factors shown.

fiable on the echo sounder and which are found on a chart. The features must, of course, be well charted. It is best to use large- or medium-scale charts, although this will depend on the individual feature. Often it is possible to locate the feature seen on the echogram upon a small-scale chart, but its details may be lost at a 100-fathom interval. Hence it would be difficult to pick off position on the chart to exacting coordinates if one wanted ±25 to 50 yards, even assuming the feature could have been charted and compiled at these accuracies.

Positioning by canyons

Crossings of canyons are almost always recognized on the echogram, and many canyons are extremely well charted. A line of position can be obtained during the crossing, with the maximum depth seen compared with the chart. Although the axis of a canyon is a linear feature, it is not always straight; however, since it descends almost continuously overall, the one crossing may also suffice for the fix. Several crossings suitably spaced will verify position. The features must be analyzed carefully because of incongruities along its lengths. On larger scale charts it will be noted that some canyons waver in their descent rate, with the downward trend interrupted for intervals, then resumed. It is not to be supposed, either, that canyons are always found in otherwise flat areas of topography, and this alone can make for difficulty in initial detection. The crossing, when made normal to the linear trend, yields the profile with maximum depth at the smallest compressed scale. This would be the actual perpendicular profile, taking into account exaggeration of the echogram. When an oblique crossing is made, it is more difficult to correlate the data with successive crossings, although the maximum depth itself would be identical to that recorded at the same point with an exact normal crossing.

Positioning by seamount

There is no single correct way to utilize canyons for positioning. Similarly, there are no formal rules for correlating observed seamounts, trenches, ridges, etc. Isolated seamounts serve as unique points without the possibility of confusion with other seamounts, but an isolated feature may be more difficult to detect. Groups of seamounts are more easily found through standard dead reckoning (DR) methods with the different minimum depth identifying each. If one sees on the echogram a seamount with its minimum

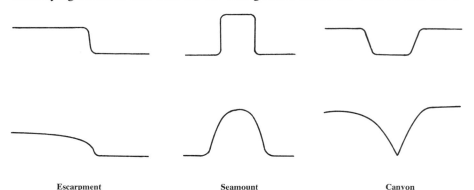

Figure 52. Representative classes of bottom-feature design. The features depicted on top will be represented on echograms in the shapes and forms shown below because of the width of the transducer sound cone and travel time of the ship.

depth identical to that of one upon the chart, the assumption may be made that the ship has passed directly over the top of that seamount. The difference in minimum depths is an indication of how far to the side the ship has passed, when it is certain that identical features are being compared. This will be discussed further under the Side-Echo Technique.

Display effects of wide sound cone

The characteristics of a wide sound cone cause representative classes of design to appear on the echogram, such as canyons or trenches appearing shoaler than their true depth. If an escarpment is symbolized by a perpendicular drop, a seamount as a bar, and a canyon as a box depression, these will show on the echogram as illustrated in figure 52.

Procedural techniques

The procedural techniques in table seven cover profile matching, contour advancement, line of soundings, and side echo, and yield accuracies (here used to mean repeatability) of a high order. They are strongly dependent upon the availability of proper reference data, i.e., accurate and well-contoured bathymetric charts at proper scales.

Profile-matching technique

The profile-matching technique is what its name implies; it is a set of procedures designed to match and compare observed profiles with sets of reference profiles. The reference profiles are obtained at some time in the past with accurate electronic control. The concept of correlation is valid because the recovering ship uses the same equipment and steams at the same speed on which the original profiles are based. Given these circumstances—sets of reference profiles precisely obtained, and use of common equipment and speeds (and direction, of course)—excellent correlation can be achieved. From this correlation, accurate estimates of position are taken.

It would seem to the advantage of the recovery ship to attempt to achieve a bottom profile identical to one in the reference set. But the success of the technique rests upon ability to interpret observed profiles that fall between any two of the reference profiles. Frequently, a more accurate position can be obtained if one can determine that an observed profile falls exactly halfway between two reference profiles. It would be rare if a reference profile could be duplicated exactly, and it is these subtle differences that account for accurate position determination.

A large measure of importance rests with choice of the area in developing the reference profiles. Typically, this is done in the following way: an area of the bottom is chosen approximately five nautical miles on the side. The exact dimensions are predicated on the number of unique reference profiles that can be obtained over the kind of topography seen. The topography should be of a nature that allows distinction among the profiles in different portions of the area; yet, it must be of a kind which allows interpretation within the area. Very complex topography is unsuitable; so is a very flat area. The reference profiles are obtained at given speeds with given equipment in a prescribed direction. Two sets of reference profiles are obtained, each at a right angle to the other, at intervals which may range from 500 yards to 2,000 or 2,500 yards. For convenience, cardinal points are used. The recovery ship enters the area at a cardinal-point direction, and the profile attained is compared with the reference set. This is a simplified description, but basically the technique works just in this way.

Figure 53 depicts a fictional area appropriate for this purpose. Two distinctive characteristics in addition to the slight rise in the northwest portion are seen; a small extended ridge trends ESE from the rise, and a very slight bowl-like low trends in the same direction from a point south of the rise. The area may have been chosen for these characteristics, and it remains to observe the profiles before judgment can be made on whether it was a good choice. The survey ship obtains the numbered profiles on the tracks shown, in this case

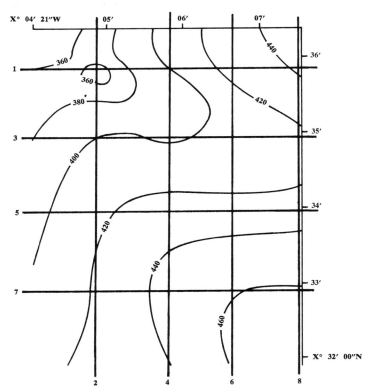

Figure 53. Fictional area for profile matching. The lines indicate North-South and East-West profiles over an area chosen for the profile-matching method of positioning.

approximately 2,000 yards apart. Profiles are obtained in both directions. These are afterwards plotted in sequence, as shown in figure 54 for the East-West profiles and figure 55 for the North-South profiles. The recovery ship—weeks, month, or years later—attempts recovery by comparing its observed profiles with the plotted reference profiles.

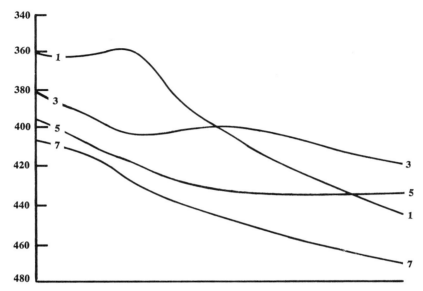

Figure 54. East-West profiles aligned in position and plotted in sequence.

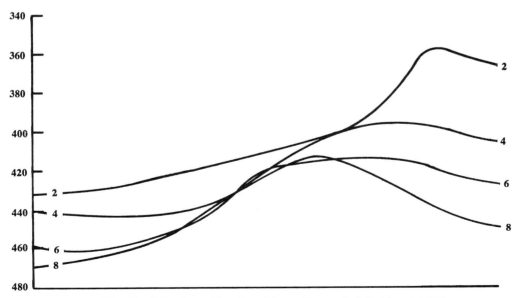

Figure 55. North-South profiles aligned in position and plotted in sequence.

When plotted in sequence but in correct relation to one another, the profiles are sometimes seen to intersect. Such portrayal is a function of the relief. For convenience these can be rearranged and plotted so each is clear of the other, as in figure 56 for the East-West profiles and figure 57 for the North-South profiles.

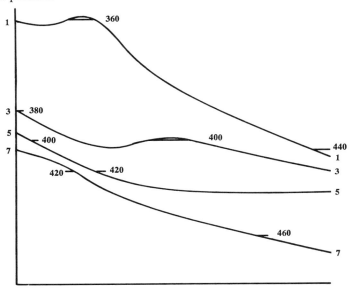

Figure 56. East-West profiles aligned so no intersection of lines is seen.

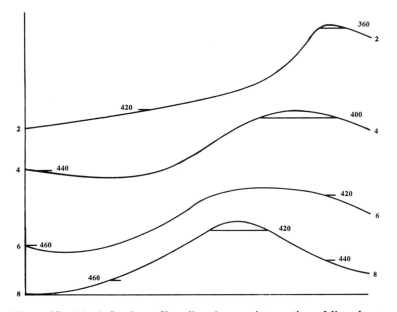

Figure 57. North-South profiles aligned so no intersection of lines is seen.

If the area for reference is carefully chosen, the resultant set of profiles will be integrated into a series that allows differentiation to a remarkable degree. It is not necessary for the ship to steam through the entire area before comparing the profiles. Often it is sufficient that beginnings of a profile are seen; if that profile is at a depth amenable to interpolation between two reference profiles, recognition and position can be instantaneous. The reverse is sometimes true, and it is possible that the observed profile cannot be recognized even after the ship has passed through the area.

Assume the recovery ship records the profiles seen in figure 58. The first is a virtual duplication of one of the reference profiles (profile three in figures 54 and 56) and, in fact, cannot be distinguished from it. If the recovered profile were slightly different, the problem would be one of ascertaining on which side of the reference profile it fell. The second profile in figure 58 falls logically between profiles five and seven in figures 54 and 56. But can it be assumed to fall halfway between? No, in this case, as the differences in minimum depths indicate slightly closer proximity to reference profile five. In the third instance, the observed profile is in the direction (N-S) at a right angle to the others. Perhaps this can be left to the reader to interpret and position relative to the reference set.

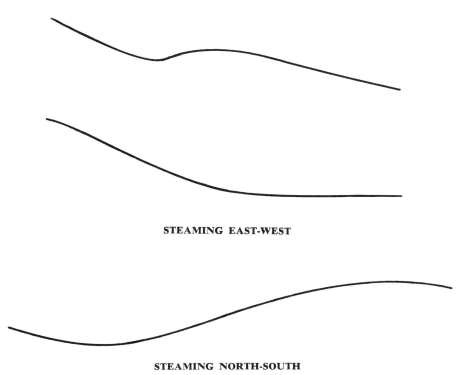

STEAMING EAST-WEST

STEAMING NORTH-SOUTH

Figure 58. Profiles recorded by the recovery ship to be compared with the reference profiles of Figures 54 and 56.

Side echoes are not entirely unwanted in use of this technique, although they are not necessarily sought. As the ship proceeds to enter an area, a unique line of position can be seen if the depths so indicate. The objective is to identify the line of position while still in the area, so the fix may be obtained through speed-time relations. When the ship is halfway through the area, the latitude and longitude can thus be picked off the chart.

Greater accuracy in relating the position of the observed profile geographically can be obtained by annotating the initial plot (figure 53) in minutes and seconds. Use of a transparent overlay on which the observed profile is traced directly from the echo sounder can also facilitate the comparison process. If there were no time limitations, the user could verify position in other ways. The most expeditious way would be to come about on a reciprocal course and head—not for a line duplicating the observed profile—but for a point splitting the tentative position of that profile and the next referenced one in sequence. With a view to greater accuracy, one might question the line spacing of the original data. If 2,000-yard spacing is good, would not 500-yard spacing yield more accurate profiles? Two factors determine optimum line spacing in the original development. One is *control*. In the absence of very good control, spacing closer than a given value is not feasible. The other factor is the *relief* in the area. A certain topographic arrangement would allow for closer line spacing (to a point limited by control compatibility) without loss of identity. In more rugged areas less profiles may be required, and in flat areas, more profiles. This is not always true, however, and the better solution is to choose an area neither too complex nor too devoid of relief.

Contour advancement

Contour advancement is also appropriately named, since it is descriptive of the technique. No bathymetric anomalies are required, such as seamounts, canyons, or ridges; but some displacement of relief is necessary. In this case it is the slopes that are of interest. Ideally, slopes of greater than one degree, but no greater than four or five degrees, are required. The slopes should not be constant because the linear distance between contours on the chart should not be equal. The technique allows for accuracy of position to be achieved on the order shown in table six, a repeatability of about 100 yards. Analysis of results indicates that this accuracy is approximately one-fourth of the line spacing of the survey producing the reference chart.

In noting that some slope is required, the presence or absence of contours with the scale of chart used must be kept in mind. Areas that appear devoid of contours and apparently flat on one chart may, upon use of a large-scale chart, show some relief or slope. In assuming that a given area is not absolutely level, use of contour advancement is facilitated by use of the largest scale chart available. Again, the chart must have been properly surveyed and compiled.

Refer to figure 59 as the base chart. The DR track is shown as a dashed line, with observed depths marked off at increments equal to the charted contour interval. These data are obtained previously (while the ship was steaming to the area in the same direction) using the 700-fathom curve as reference contour. This is a time-distance plot and is based on ship speed and recorded depths. For example, if the ship had been steaming in the same direction for several miles (over this slope), it would be recording incremental depths differing in time from recording of the charted depths. The recorded contour crossings are merely extended to the area on this chart and plotted in relative position on the DR track.

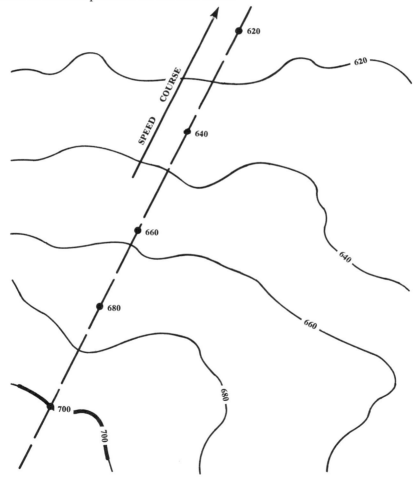

Figure 59. Base chart for contour advancement. The DR track (dashed line) has marked upon it recorded depth in the same interval as contours on the chart.

By starting with the 700-fathom curve as the first one crossed, the curve is traced onto an overlay and becomes the *reference contour*. The assumed track is also indicated on the overlay. The next step is to shift the overlay in the direction of travel until the reference contour (700 fathoms) matches the plotted (not contoured) position of the next depth for which a contour exists (680 fathoms). Now the 680-fathom curve is traced. It should intersect the reference contour at one or more positions. The overlay is again advanced along the direction of travel to the point where the reference contour intersects the plotted 660-fathom curve. This contour (660 fathoms) is now traced. The three contours that are now traced should intersect at a point off the assumed track. Or, some triangle of error will be indicated which can be further defined by continuation of the "advancement." The intersection of lines becomes the position which is used to adjust ship's track. This adjustment is a linear shift of the DR track and is a fix modified by the time (and therefore positional) lag after determination.

This description may appear complex in the absence of an overlay. Actually, the technique is a simple one and easy to master. Figure 60 represents the overlay after the contours have been traced by advancement of the 700-fathom curve.

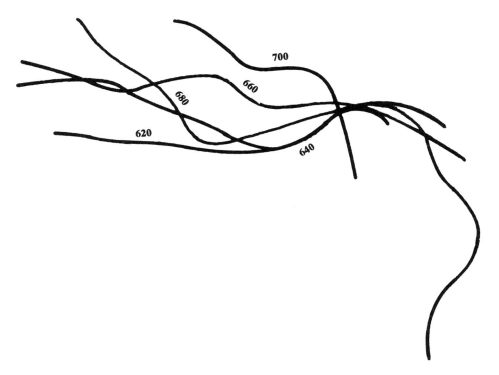

Figure 60. Transparent overlay. The overlay shows the contour traced successively, and the intersection shows the position of the ship.

DETERMINATION AND RECOVERY OF POSITION 87

Line-of-soundings technique

The *line-of-soundings* technique is not definitive in form and structure. It is procedural because the individual must accomplish certain steps, but its proper use is not dependent upon formal procedures followed sequentially. Characteristically, the technique of line of sounding leans heavily on accuracy of the reference chart. Consequently, there is direct relation between the reference chart and accuracy of the position finally determined. This technique is less dependent on scale of the chart chosen as reference, other than probable greater accuracy of larger scale charts in certain areas.

Figure 61 shows the DR track with contour crossings marked as recorded and at the same contour interval as on the chart. The other parallel courses are the possible correct tracks assuming the DR track itself is incorrect. It is incorrect in this example, as seen by the discrepancy in crossings. When the crossings are plotted upon an overlay, this can be moved until a proper match is found. The overlay is now aligned, probably parallel to the DR track and perhaps fore or aft of one of the positions.

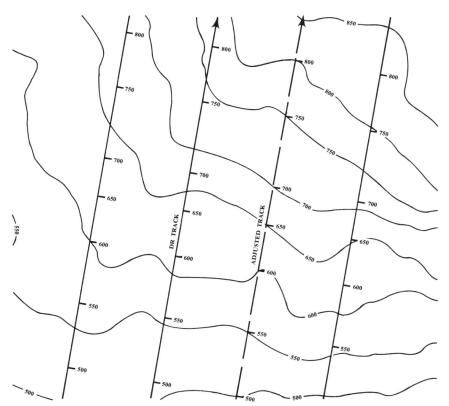

Figure 61. DR track for line-of-sounding technique. The illustration depicts the adjustment of track (marked with recorded depths) to match known contours on chart.

Only a line of position can be achieved if the contour crossings are at the same interval. This occurs because equally spaced contours allow movement of the representative crossing in one direction alone. They would "fit" anywhere in the other direction. Thus, this technique cannot be used with constant slopes. Line of sounding is very similar to a technique used for many years, called the "onion skin" method. Essentially the same concept was applied. Onion skin paper was used in the days before clear acetate or some other similar material came into routine use on ships.

Side-echo technique

The *side-echo technique* is used with seamounts. The phrase side echo is in this case a misnomer and actually concerns position to a side of the features instead of the phenomenon which the term usually defines. Conceptually, the technique depends on the intersection of lines of minimum depths recorded on at least two passes over a seamount. The point of intersection is the position of the seamount top in relation to the ship.

Although the procedures are not difficult, several important concepts must be kept in mind. Seamounts (or seaknolls) are large features. The notion that they are steepsided, as seen on echograms with large vertical exaggerations, must be dispelled. The tops are never so sharp or well defined as they appear on the echogram. Charted values indicating a unique minimum depth can be incorrect; the depth can be wrong; or (more probably) it may have been charted in the wrong position. Knowledge of whether the feature was compiled from a survey or from random tracks is essential for confidence in use of the reference chart. One records the minimum sounding of a seamount identical to the depth on the chart when the ship passes directly over the top or very close to it. A recorded value shoaler than seen on the chart indicates that the chart is in error (or the sounding gear is in error or it is the wrong seamount). It is desirable to know the nature of the equipment used by the ship(s) that produced the data upon which the chart is based, and of course one must know the limitations of his own equipment. All these must be kept in mind. There is also the possibility of using incompatible features, and an isolated seamount is best chosen despite the greater difficulty in initial detection.

The following discussion refers to figure 62. Once the seamount that is to be used is selected, an approach from within 20 to 30 nautical miles is made. The DR track is extended to cross the charted position of the seamount's minimum-depth. Starting position will vary with local conditions, but in calm seas the DR track can be made good from this distance. It is unlikely the feature will be missed altogether if it is isolated and of any extent horizontally. Final approach should be adjusted as indicated necessary by recorded bathymetric data or other means. As the ship approaches the top of the seamount for the final run, constant course should be maintained.

Observed depths are noted as the bottom rises. These values are taken at given time intervals, perhaps every minute, and transferred to the DR track (or upon an overlay to the track). The minimum depth is noted and so marked; this is 1,138 fathoms in figure 62. When depths are seen to decrease for a time and with values appropriate to depiction of the feature on the chart, it can be certain the seamount top has been passed. The marked depth values will coincide altogether only rarely with charted contours.

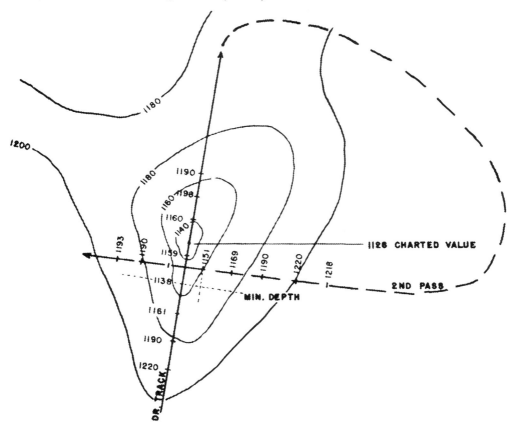

Figure 62. Relationship of tracks over seamount for use with side-echo method.

After the ship has passed the top, one of three things must have occurred:

1. The ship passed *directly over* the top or very close to it. This is immediately apparent if an observed minimum depth of 1,126 fathoms ($\pm$ two to three fathoms, to account for charted and systematic errors) is seen. Since ship's speed and course are known, the ship's position can be determined to a high degree of accuracy because it has passed over a known geographic point. No further action, other than routine verification, may be necessary.

2. The ship passed *to one side* of the top at a distance where the top lay outside the area covered by the beam width of the echo sounder. This is apparent if the navigator observes a minimum depth greater than 1,300 fathoms; this means that the ship was more than 1,300 yards to the side of the top. For a 60-degree sound cone the geometry of the situation leads to this rule of thumb: the deepest value recorded with the seamount top still retained within the sound cone is twice the distance from ship to seamount top horizontally. Obviously it is to the advantage of the navigator to choose, if a choice exists, a seamount with deep minimum depth, other considerations being equal. In the event that the top did not fall within the sound cone, the ship's position can be determined only approximately. During actual attempt the ship would then return on a reciprocal course to attempt recovery once again.

3. The ship passed *not directly over* the seamount top, but *close enough* that the top fell within the sound cone. This is seen by observing that the recorded minimum depth is between 1,126 and 1,300 fathoms. The side-echo technique of positioning applies to this situation.

Again referring to figure 62, assume that the minimum depth recorded was 1,138 fathoms. Locate the position of this depth on the track and draw a line at a right angle to the track passing through the point of minimum depth. The line is extended on either side of the track, as it is unknown at this time on which side of the seamount top the ship passed. The line may be labeled *minimum depth*. The seamount top lies on this line at some unknown distance and direction from the DR track. Once depths begin to decrease again and it is certain that the top has been passed, course is altered. The new course should intersect the original track (from either direction) at a 90-degree angle or as close to normal as possible. Alteration should be planned to intersect the first track with the charted value of the seamount top by change in course of 270 degrees.

The reason for having the second track intersect both the charted value and the original DR is that one did not know to which side the top had been passed. If indication of direction of the top can be available, the second course would be set to intersect an appropriate point off the charted minimum-depth value. This is mentioned because of advances in side-scanning sonar. Increased range of such devices could indicate direction of the top to the side of the track as the ship passes it the first time.

As the ship approaches the seamount top again, noted by rapid shoaling on the echo sounder, depths are marked off as before upon the new track. The point of minimum depth is again highlighted once the top has been passed. A minimum depth of 1,151 fathoms is recorded on the second run. A second right angle is drawn; this is perpendicular to the new track and passes through the point of minimum depth. The intersection of the two lines of minimum depths locates the seamount top relative to the ship. Direction and distance separating the intersection of the two lines of minimum depth are the *offset* of the ship. Actual track adjustment may then be readily accomplished by shift of recorded times and soundings by direction and distance of this offset. A third track may be chosen if necessary, and the data similarly analyzed.

Computer application to surveying

Mention has been made of computerized processes. The use of automated technology will have wide applicability for bathymetric navigation. One of the difficulties inherent in analysis of large amounts of bathymetric data is ambiguity of information. Time is ordinarily not of greatest importance; one ordinarily has time to judge data prior to coming to some decision if the data were clear cut and capable of resulting in substantial conclusions. Computers offer this advantage to the navigator—they perform statistical analysis impossible to accomplish otherwise. In the absence of computer storage and selective retrieval capabilities, the techniques for positioning and recovery just described can be arduous.

CHAPTER VI

Computer Applications and Bathymetry

Computer applications in this chapter concern use of automatic data techniques and equipment, including computers, in processing information for data handling, chart production, and navigation. Concerted efforts in the integration of data-processing concepts in data handling, chart production, and navigation are being made at the present time. Reference should be made to any standard computer technology manual for explanation of terms used in this chapter.

Data handling

A whole new field of literature is devoted to automatic-data processing and use of its component equipment, digitization of data, and general application of computers to the overall problem of data handling. Data handling is a general term which includes the collection, storage, retrieval, display, and other manipulative steps necessary to transform raw data into some usable product. Most of the readily available information on this subject is highly technical, and relatively little is found on echo-sounding application. Government and private oceanographic institutions have published accounts of bathymetric automation from the data-processing view, but by and large the automation of depth information has lagged behind such application to other oceanographic variables.

Although computers have developed to the point where even routine aspects of our lives are affected by them, it must be understood at the outset that computers are incapable of performing any function which they have not been programmed to do. If the function involves choices to be made, that choice made by the computer is predetermined by what it has been programmed to do and to accept under certain circumstances. There is, of course, the greatest dependency on validity of the information on which these choices are based. Fundamentally, the advantages of computers are threefold—ability to store tremendous amounts of information, to draw upon any part of it and do with it exactly as directed, and to present it in any form desired very rapidly and accurately.

Automation of bathymetric data collection

Automation of bathymetric data collection proceeded along three lines of development. The first approach was application of available automatic data-processing equipment to standard echo sounders; the second, of new types of sounding transducers in multiple-mode systems; and last, of *swath sounding* to obtain closely spaced profiles perpendicular to ship's direction.

Early application

The first approach to bathymetric data collection was application of available automatic recording equipment to standard echo sounders, or modification of such equipment to accept the depth input on tapes or punched cards. There was nothing inherently difficult in doing this, and the necessary information was recorded, stored, selectively retrieved, and used for further manipulation. The quantitative amount of information was of any magnitude desired, depending on the particular equipment used. One needs to know the depth value and latitude and longitude. The control variables are vital, too, such as rates, time of day, stations, and error standards. And it is desirable to know and store data on sound velocity or the factors that determine it, subpenetration data, kind of echo sounder used, coordinates of track with start and finish, changes in course along given lengths, name of the ship, ship type, draft, inputs from radio WWV, etc.

As is to be expected, the Navy is in the forefront of those seeking to automate bathymetric data collection. Within the community of interests that are involved in bathymetric work, the major endeavor is probably occuring in or for the Navy. That many of the advances are being accomplished by the Navy is not surprising, considering the impetus and the mission involved, yet this is somewhat ironic since the Navy was at first slow in adapting to the concept.

Use of magnetic tape to collect, store, and manipulate deep-sea soundings was first proposed formally within the Navy in the mid-fifties and discussed even before then, yet nothing was done because projected benefits were deemed not to be in proportion to the cost of installing recording equipment. Difficulties were forecast because of uncertainty of the permanence of magnetic tape and even for the lack of specific requirements against which to apply the process. At any rate, the Navy, finally taking account of the direction in which automation was invariably moving, soon applied efforts to develop automatic depth-collection equipment and office-processing techniques. These were, in part, independent attacks on depth collection alone and, to a larger extent, integration of echo sounding into larger systems collecting general oceanographic data. The active aid and support of industry were sought, and probably the most unique concept that evolved—conformal echo-sounding arrays, or "swath sounding"—was first proposed by industry. The Navy is currently advancing technological development of sounding techniques and data processing at a fair pace, both through its own direct work and by management of associated contractual efforts.

Automation and multiple-mode systems

A second line of development for bathymetric data collection considered use of new types of sounding transducers in multiple-mode systems. Older methods of achieving directivity and stabilization required a very large transducer, whose diameter was up to 90 cm. Gyro-driven potentiometers provided stabilization by controlling servo-amplifiers which in turn directed hydraulically activated pistons; one potentiometer compensated for pitch and another for roll. This mechanical process often failed, and the transducer became cocked to one side of the vertical, sometimes without the operator being aware of it.

The new method of electronic stabilization and directivity avoids this difficulty and can be achieved in several ways, one of which is to use separate transducers producing preformed beams, perpendicular to one another. The outgoing beam may be stabilized for pitch, and the receiving beam for roll. The mode of operation may include use of towed transducers and concurrent directional and nondirectional utilization at frequencies sufficiently different to be recorded separately. This is usually part of a multiple-mode system, with the bathymetric subsystem being part of a larger oceanographic collection system. Use of processing equipment is facilitated because of storage capability for more than one kind of variable, and an optimum configuration of equipment is available under varying collection conditions. The concept bears out the view that once an expensive platform is in an area it should be capable of collecting every possible oceanographic variable.

Swath sounding

The third line of development for bathymetric data collection is a conceptual one, using both new echo-sounding equipment and new techniques. This has resulted in novel forms of transducer arrangements with the following characteristics; they are highly directional, sweep a path approximately amidships and at right angles to the ship's track through electronic interplay of conformal arrays of many transducers, and yield bathymetric data over a "swath" on either side of the ship's track. Figure 63 illustrates how the swath-sounding method obtains a spread of data at a right angle to the ship's direction of travel. Figure 64 is a bird's-eye view of the swath-sounding method shown here as completed contours behind the ship.

Since the swath-sounding system obtains a spread of data on either side of the ship's track, it can be thought of as stripping away widths of surface

96   BATHYMETRIC NAVIGATION AND CHARTING

layers to reveal what lies hidden below. If one pictures the bottom as being contoured (at the interval chosen and the scale allowable by the system), then the ship's tracks peel away layers of the surface hitherto unseen. Or, picture an actual set of contours at desired scale on a large sheet of paper. Covering these contours are inch-wide strips of opaque tape in many directions. If the strips of tape were removed, a slowly emerging set of contours would result. So it is with this sounding system. Each ship's track lifts one

Figure 63. Swath-sounding method. This survey system uses multiple-beam sonars which give not only a depth beneath the survey ship, but also many additional depths on both sides of the ship along a line perpendicular to its track. With associated computers and plotters, strip contour charts can be produced in real time from a single survey track.

COMPUTER APPLICATIONS AND BATHYMETRY 97

more strip of tape, irrespective of direction of travel. Many areas would be duplicated unless systematic coverage had been planned, and some areas would remain concealed unless a ship passed over them. But slowly and inexorably a pattern must emerge, with the pieces remaining uncovered to be surveyed at some other time. The ship computer can store the tapes of any track and can quickly play any of these into finished contoured products.

As to width of the swath when compared to the standard wide-beam coverage, comparisons are not direct unless differences in the survey approach are kept in mind. Figure 65 illustrates the relationship of standard transducers, both wide-beam and directional, to the swath system. This should be looked at carefully, recalling that wide-beam transducers are fixed to the keel of the ship and accordingly may cover an area greater than its normal 50° to 55° cone allows. Conversely, the ability to record swath profiles from 45° on either side of the vertical is, in practice, modified because of refraction errors and other difficulties. Even these distinctions are not directly comparable because of the multitude of directional profiles that the swath system records at a right angle to the track and because of its applicability to automation.

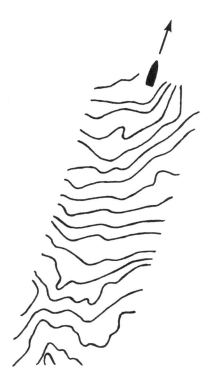

Figure 64. Swath of data obtained as ship runs a line. A bird's-eye view reveals a swath of bottom depth information shown here as completed contours.

Recapitulation of data handling

To recapitulate the directions taken in automation of the data-collection or echo-sounding phases, the first attempts were limited to storage of sounding data on magnetic tape or punched cards. Advancement to a systems approach occurred naturally, with depth information recorded, stored, and manipulated as part of a larger amount of oceanographic data so handled. Finally, what is here called a "swath sounding" system evolved out of computer technology and without recourse for dependency on traditional equipment or techniques. Table eight lists miscellaneous data applicable to the different sounding means in use.

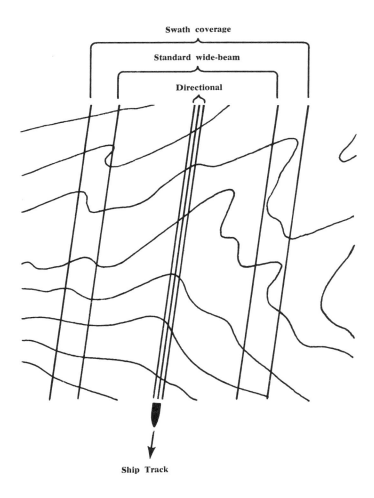

Figure 65. Relationship of standard transducers, wide-beam and directional, to the swath system.

TABLE 8.

Pertinent data on different sounding methods.

	Concept	Profile	Effect	Automated	Applicability to Chart Production
Standard Sounding Equipment	Pt to Pt Sounding	Longitudinal	Composite	No	Meager
Automated Standard Sounding	"	"	"	Slightly	Meager
Mechanical [1] Stabilization	"	"	Less Composite	"	"
Electronic [2] Directivity	"	"	"	"	"
Systems Approach [3]	"	"	"	Yes	Greater
Swath Sounding	Mid-ship Sweep	Normal to Track	Highly Directional	Highly	Chart Produced Automatically

[1] With large transducer heads and increase in operating frequency.
[2] Stabilization of preformed beam achieved electronically.
[3] Inclusion of varying modes of operation as part of larger survey system.

The computer can be programmed to generate a host of other valuable information, including a single-point profile of the track itself. Concurrently with swath sounding has come computer manipulation of the data, resulting in automatic correction for refraction and other errors, control input, selective intermittent storage, processing of the information into lists or tables, and computer-produced contoured charts. When carried through to the ultimate, the system provides for automatic production of charts in real time.

Future computers

In the future, computers will be reduced in size and cost through use of integrated circuits and modular units. Self-heating circuitry will result from aerospace development. New computers will be developed using light techniques. This will depend on rate of development of lasers, fiber optics, and photodiodes. Card speeds will be increased to upwards of 5,000 per minute utilizing fiber optics sensing. Magnetic-tape storage capacities will increase, and printers will benefit from thermal and laser research. The product of display consoles will be available in color, and speech-operated consoles will be in limited use. Computer-controlled closed-loop systems will seek an image, translate it to digital form, and transfer it for storage to another computer. Computer memory will increase, and access time will decrease. Meantime failure of several thousand hours will be achieved.

In the application of software, families of computers will be available using the same basic instruction sets, character codes, word sizes, and formats. Microprogrammed circuitry will enable one computer to execute programs prepared in the machine language of another. It is also obvious that standardization of all phases of software will increase. Systems of computer usage will evolve along the lines of large central computers serving independent users from remote terminals, networks of smaller computers sharing data banks, and mission-oriented computers tailored to specific functions. Finally, many new areas of application will emerge because of the availability of powerful computers at low cost and development of unique user-oriented systems languages.

Chart production

Even now, relatively few survey or chart production techniques are automated to the point consistent with present state-of-the-art capability. The automated chart construction techniques that have been developed so far are primarily cartographic in nature and do not pertain to bathymetric or nautical chart production alone. Moreover, they have occurred generally as side effects from efforts of the purely cartographic agencies of the Federal Government. Since approximately 1962, three cartographic and nautical functions of production have been automated: construction of chart projections, plotting of electronic navigational curves, and placement of sounding depths. Even so, manual accomplishment of all of these functions has not entirely disappeared. The great bulk of work necessary to produce a chart is not automated at this time; this includes assimulation and evaluation of source materials, compilation and scribing of color manuscripts, and, to some extent, selection and placement of type. Efforts are now being made to automate these processes. This entails new equipment and software programs requiring some systems approaches, with assistance to the governmental cartographic agencies by industrial contractors.

Data analysis

The problems attendant to automation of bathymetric data assembly and analysis are probably unique to organizations such as the Naval Oceanographic Office. Probably no other kind of institution receives and processes this great amount of deep-sea soundings on a worldwide and all-source basis. It is here that ambitious efforts in automation of bathymetric data analysis are being applied. Complex as this particular undertaking is, it is but one facet of bathymetry. The organization is involved also in establishment of bathymetric collection requirements and the conduct of bathymetric surveys for the Navy, as well as developmental work in nautical data processing in general.

Sources of bathymetric data

There are two sources of bathymetric data: documents transmitted by Navy, Coast Guard, Merchant Marine, certain foreign merchant ships, and scientific ships, on the one hand; and survey data on the other. The former group consists mostly of smaller scale track charts, with fixes, time and soundings approximately every 15 minutes, log sheets which depict similar information, and annotated echogram rolls. The volume of survey data compares with random data on the order of 1:40. Survey data is usually of smaller ocean areas and on larger scale charts, and almost always of higher quality than the random data. Although the volume of survey data is smaller, the consistently high quality makes it highly valuable. Reasons for the difference in quality are easily understood—the survey data are obtained by trained personnel using precision echo-sounding and control equipment.

Processing of data

Some, if not all, of the preliminary processing is done in the field, or in the office by similarly trained personnel. The base chart used for field plotting of random tracks is the 3,000 Series Plotting Sheets with contoured overlays at 100-fathom intervals. The scale of this series varies with latitude—about 1:400,000 to 1:1,350,000 with four inches on the chart representing one degree of longitude. A typical document might entail five to eight plotting sheets. In the processing stage the depth data are used in plots on master sheets, after which the plotting sheets are microfilmed. Analysis and acceptance or rejection of the information are a collective endeavor, depending on the echogram, the plotting sheet, and other known information. The index for this series of plotting sheets in the North Pacific is shown in figure 66.

Production of master sheet

Since many random and survey tracks covering the same area are available, conflicting information is frequently noted. Once the validity of the sounding data is determined perhaps with recourse to the echogram for the shape and depth or to original logs for position, a master sheet is maintained by plotting-sheet area. This master sheet is an all-source collection of analyzed and validated depths, plotted geographically in correct relation to each other. These sheets are then contoured at a 100-fathom interval, incorporating known survey data as applicable. (Obviously this description is a much abbreviated version of an intricate process.) Output is in the form of many specialized products in addition to standard contoured plotting sheets, among them bathymetric charts for engineering purposes, special-area analysis charts, beach-landing charts, and cable-laying charts.

Automation of chart production

By far the major portion of this work is applicable to automation. Statistical weighting can accommodate classes of soundings, and microfilming can reduce, to reasonable proportions, storage loads of original data held. Finally, computer programs that automatically draw contours from a taped or punch-card input are feasible. The bathymetric data thus engendered draw upon only a small part of computer memory devices and could be recalled at will.

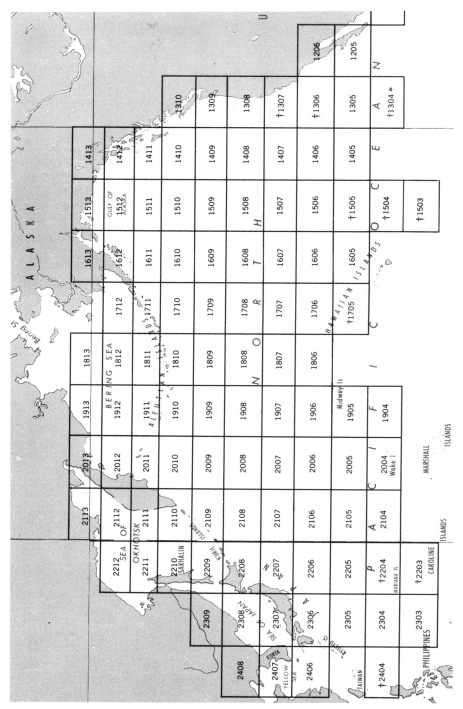

Figure 66. BC plotting-sheet index of the North Pacific.

Digitized sounding data could form part of a central holding index, including all known data over given areas, evaluated and subject to recall, display, and use in production by amounts or categories. More than one set of sounding values per given area could even be stored, and analysts could query the computer from remote stations. If this system were used, depth anomalies that differ not topographically but statistically from stored data could be instantaneously known. Analysis of new data would be accomplished quickly, and the new data inserted in the central holdings.

Not all of these methods can now be accomplished through automation, but they are desirable and probably will be done shortly. Some are being done now. Should such a total system become operational in the way described, it would not be difficult to foresee imaginative use of the data for bathymetric navigation purposes.

Computer applications for navigation

Most of the techniques for bathymetric navigation, such as profile matching and line of soundings, described in chapter five, can be automated. Certain of them are particularly adaptable to computer application, especially those that require recognition and comparison. Of the techniques whose nature is chiefly procedural, profile matching and line of soundings involve processes well within capabilities of newer shipboard computers.

Profile matching and line of soundings

With the profile-matching techniques, for example, as a ship begins to traverse a reference area, the computer may be programmed to begin comparison at that moment and continue yielding navigational information. This would be updated as more of the observed profile is recorded and analyzed, with the computer firming on its first statistical choice or refining its estimate in real time as necessary.

The line-of-soundings technique is merely a best-fit analysis with obvious applicability to data storage and comparison. So it is, in dealing with techniques where all or part of the positioning process can be automated, that some computer output can be generated. This output is a conclusion on the part of the computer based on its programmed instructions, and the navigator must then make a judgment taking this conclusion into consideration. The individual will rarely dispute these results, yet it is recalled that navigation is an art. The computer is merely a tool, and to be used well it must be used wisely.

Bathymetric and space navigation

The most spectacular use of computers for navigation was demonstrated to the world late in July 1969 when the United States achieved the first moon landing. Every phase of the historic journey: blast-off, earth-parking orbit, translunar trajectory, lunar orbit, return-to-earth, orbit, and re-entry, involved intricate mathematics. Computers provided the near-instantaneous solutions which permitted exact timing and precise control necessary to successful completion of the mission.

Tracking and communication coverage of all the possible maneuvers were impossible with only land-based stations, and the National Aeronautics and Space Administration utilized precisely positioned instrumentation ships to support the extensive control system ashore. One of the means of insuring absolutely correct positioning of the instrumentation ships was bathymetric in character. In a sense, this was a redundancy, as other backup systems were available; the bathymetric position was an interim step used to reset an inertial navigator. Such precise bathymetric positioning is desired nonetheless. This is achieved through an approach similar to a numerical topographic mapping technique developed at the Army Topographic Command of the Corps of Engineers. In effect, a matrix of X, Y, and Z information (latitude, longitude, and depth) is collected through a survey and stored in a computer. Programs are written to direct the computer to generate a linear strip of profile data upon command. The computer may be told that a ship enters the (previously surveyed) area at given coordinates and traverses it at a certain azimuth. It will then generate the profile most likely to be achieved over the bottom if the ship makes good this track.

Difficulties encountered are identical to those inherent in the data acquisition phase for use with any bathymetric positioning technique. Firstly, the data collected must be based on the best control available. Secondly, the survey areas must be carefully chosen so as to present topography that is subject to minimum misinterpretation. Differences in equipment compatibility are eliminated because the ship positioning itself will be the one that collected the original survey data. This may be an important consideration in choice of area made, as certain topographic configurations necessitate identical equipment for survey and location phases. In other cases, recognition is possible using different equipment.

The chart and the computer

In its memory bank the computer can store any information shown upon a chart. This is not to imply that it will replace the chart; there is no information on a chart, however, that cannot be computerized. This includes not only depths and such items as lists of lights or buoys, but also contours, shoreline delineation, lines of electronic control, and magnetic deviation. All such information could be stored, but available for instant recovery and display in any form desired, and could be displayed graphically. To consider it in this sense, the chart actually becomes one of the by-products of the computer. The other outputs would be the things one ordinarily does with a chart: distance measurements, study of bottom configuration, choice of navigable channels, etc. Here the work would not be done upon the chart but would be done by the computer. What this further allows are statistical queries and answers, e.g., "How many seamounts in this area are greater than 400-fathoms minimum depth and extend overall to 2,200 fathoms?" or "How

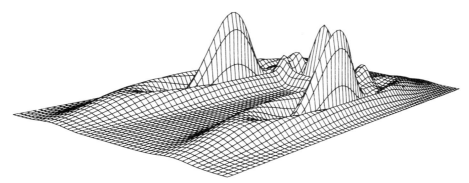

Courtesy of California Computer Products, Inc., Anaheim, California

Figure 67. Computer-generated relief drawing. This technique can depict relief from different perspectives and at different vertical exaggerations.

many flat areas exist within a given portion of the chart of three-degree slope or less, and of these how many are found between 1,200 and 1,350 fathoms?" The computer could give the number of such seamounts or flat areas easily, but it could also graphically chart the items. Available printing mechanisms could also display relief in perspective (figure 67).

Future bathymetric requirements

Important trends have been established in data-processing technology which portend future direction and accomplishment. These trends must affect navigation in the broadest sense as well as in matters purely bathymetric in nature.

One may well pose a question regarding the effect of future developments on bathymetric data collection, chart production, and navigation. The significance of advances in data processing to bathymetry in all its ramifications is far reaching. It would be difficult to describe the bathymetric-oriented systems that will emerge but not difficult to predict the impact. Much of what lies in store can be gained from examination of advances to date: development of future bathymetric systems which exploit data-processing technology to the maximum will reflect as great an advance over systems presently in use as current echo sounding represents over the lead-line techniques of years ago.

Increased computer application must occur to meet known and anticipated requirements necessitating bathymetric inputs. Requirements for certain data exist even today that cannot be wholly met in the absence of automated processes. These requirements manifest themselves in specific needs for slope studies, topographic masking information, submarine and other submersible navigation data, site-selections for a host of engineering applications, cable routes, search operations, war gaming, and weapon-system effects. As this capability increases, new requirements will undoubtedly develop. This, in turn will necessitate introduction of novel data collection systems, tailored to the kind of data necessary for computerized sorting and processing. It is entirely possible that survey platforms developed to accomplish this phase of the process will replace survey ships as they now exist. Completely automated integration of collection, analysis, processing, and production will necessitate review of established echo-sounding techniques.

CHAPTER VII

Needs and Future Developments

Forecast of developments

In the years since echo sounders were introduced to general use, charting has undergone radical change. Under the impetus of technological advances in electronics and computers, and spurred by a greater need for ocean bathymetry, programs of bathymetric collection and chart production initiated only a decade ago have become largely outmoded. The importance of every aspect of oceanographic investigation is receiving somewhat belated recognition, and there is evolving better nationwide coordination of oceanographic programs.

The title of this chapter implies prediction based on needs, which, in any science subject to dynamic change, can be a hazardous undertaking. Still, some attempt will be made to outline the course of future developments in survey instruments, charts, and procedures. In some cases, discussions of future events are fairly safe, involving logical extension of what are presently conceded to be worthwhile efforts. In others, the future is unclear, and any prediction is only a guess. In most instances, however, it should be possible to make some valid forecast. Of course, a forecast can be perfectly valid yet turn out to be incorrect. In predicting the future of bathymetry, every expert is his own best prophet, and it is not probable that what is written here will meet with universal acceptance. A discussion of future developments in bathymetry has firm bearing on a more complete understanding of the navigator's job.

Dynamics of requirements

Pressing needs are not always recognized in time, but they eventually make themselves felt. It would be gratifying to be able to stipulate bathymetric chart requirements at one sitting and to know these requirements would remain unchanged for a length of time sufficient to implement means to satisfy them. But changing times bring changing requirements; this is the nature of things. Further, there are subjective factors present. While many individuals could be convinced that a requirement exists for complete world coverage of bathymetric charts at a scale of 1:250,000, it could conceivably be demonstrated that no such requirement exists. Certainly requirements do exist for accurate bathymetric charts at a suitable scale, covering many areas of the world's oceans.

Chart accuracies for various work in the oceans

Most individuals engaged in work in the marine sciences would readily concur that there is need for proper bathymetric data to aid in the performance of work/tasks in the oceans. By the term *proper* is certainly meant some reasonably correct portrayal of bottom terrain features. It is probable that many things could not be done in the absence of adequate bathymetric data. But where will these data be needed? When? What are the accuracies required? Are graphical charts the only or best means suitable? These questions come quickly to mind and are rather obvious, yet are far from unimportant. There is another group of essential unknowns characterized by even greater tenuousness, e.g., to what extent is accomplishment of given tasks affected by lack of proper bathymetric information? That is, can one quantify the element of risk per kind of task under conditions in which a deficiency of these data exists? Few persons can claim to know that a ten-million dollar program of bathymetric surveys and chart production means effective support to national goals in oceanography, while a five-million dollar program means ineffectiveness. Yet, many individuals have no difficulty in spelling out the difference between both availability and lack of charts with characteristics allowing placement and maintenance of structures on the bottom. An important intangible is absent—performance criteria for work/tasks as they may be related to chart requirements or other bathymetric input.

Criteria for satisfactory chart

What is it that would make for a satisfactory chart? One supposes the chart would be satisfactory if it could meet its intended purpose in the hands of the user. The qualities allowing that purpose to be met might be *correctness, accuracy,* and *reliability.*

To be correct, a bathymetric chart must portray the actual terrain found in a particular area. We know that this is unlikely, or, even if possible, not able to be ascertained as such. This is attested to by the difficulties in acquiring bathymetric data. Therefore, the chart is frequently incorrect.

To be accurate, the chart must be free of mistakes and errors, i.e., it must be compiled with precision; the best-possible use of available survey data must have been made. By this token, it is not likely that very many charts are accurate. Despite this definition, in common use the accuracy of a chart is deemed a standard from which deviation is measured in linear terms. Thus, a chart would be called accurate horizontally to ±2,000 meters. Even if this definition were applied, there is no way to relate properly accuracy to given portions of the chart. Would the chart be accurate to ±2,000 meters 50 percent of the time over 20 percent of the area? Few charts may be considered truly accurate by whatever definition applied.

To be reliable, the chart must be counted upon to do what is expected of it, a criterion that is more useful if described in terms defined for common use of accuracy. A chart of given accuracy is, in other words, one we may assume will be reliable, the amount of accuracy being a measure of that reliability.

Work/tasks and associated terrain-accuracy standards

Development of performance criteria by task or groups of related tasks would seem to involve no more than routine determination; however, a listing of assignments will not, for reasons which are obvious, receive general agreement for every work/task. Nevertheless, this must be attempted. Let us assume that bathymetric information is required for performance of most tasks in the oceans in the form of standard chart graphics. Table nine contains a listing of work/tasks in the oceans, together with associated terrain-accuracy standards necessary for the chart product in use.

TABLE 9.

Work/tasks with their associated terrain-accuracy standards.

Work/Task	Terrain-Accuracy Standard (Meters)
General Navigation (long-range)	±3,000
Submersible (short-range)	± 300
Submersible (long-range)	±2,000
Farming	± 200
Fisheries Research	± 400
Implantation of Gear	± 25
Gravity Research	± 50
Antisubmarine Warfare	± 50
Mine Warfare	± 50
Rescue and Salvage	± 25
Mining	± 25
Deep Drilling	± 25
Retrieval of Treasure	± 25
Man-in-the-Sea	± 20
Cable Laying	± 100
Pollution Studies	± 150
Waste Disposal	± 250

First, it must be said that no attempt is made to include all known or predictable work/tasks in the oceans. The listing is representative only. Second, some familiar aspects of work in the oceans are omitted, such as law and inspection, air-sea interaction, food from the sea, ocean engineering, etc. These efforts either are too inclusive for the purpose intended or can be divided into identifiable subtasks. The accuracies are stated in terms of terrain-accuracy standards for the chart rather than absolute accuracy for the task itself. If it is necessary, for example, to position a geodetic transponder to within ±10 meters, this may not mean (at least no one can say) that the chart used to do this must also be accurate to ±10 meters. The relation of the chart's terrain-accuracy standards to absolute positional accuracies necessary for the task may be direct, but may or may not be identical. It is true that this kind of rationale applies to certain groups of tasks only. It is believeable that chart accuracies of ±50 meters are superfluous if the requirement for submerged long-range navigation necessitates positional control of only ± one kilometer. But in the kind of task necessitating capability to return time and again to a given geodetic position, the greatest possible chart accuracy is desired.

Even supposing the data in table nine are valid, the biggest job still lies ahead. This is to translate the required terrain accuracy standards to appropriate chart implementation programs.

For general long-range navigation, positional achievement of ±3,000 meters is ample. The navigator desires to know his position to the greatest possible accuracy, but it is difficult to see how transit times could increase proportionately with positional accuracies greater than this value. True submersible navigation (excluding submarines) necessitates a thorough knowledge of bottom terrain features within a defined area per type of mission. If we assume this area to be at least 40 square kilometers, definition of features at a closely spaced contour interval should result in ±300 meters. In this and other examples, the nature of the operating terrain will affect ability to achieve such accuracies on the chart.

For farming and fisheries research, the accuracy standards seen are the author's best estimates. Many could dispute the values listed. However, recall that what are shown are the *chart standards* for the given tasks, and not necessarily those for the tasks themselves.

For the remainder of the work/tasks, there is direct relation with position and validity of measurement, as with gravity research; and sometimes the work/tasks can be augmented by visual sightings. Pollution and waste disposal studies are, of course, directly related. Here, again, the author falls back upon those more expert than he as to accuracies required. In certain cases, such as disposal of radiation wastes, exact position would seem to be critical.

Chart products for most of the work/tasks will invariably be of large scale, perhaps 1:10,000 and even larger. There is no need to compare such a chart with established accuracy standards for land maps. For example, it is commonly accepted that a 1:10,000-scale map (Class A) should have 90 percent assurance of ± five meters horizontal and ± five to eight meters vertical contour accuracies. But this product is compiled with great precision from stereoscopic photographs obtained under strict conditions.

Survey platforms

Very few systematic bathymetric surveys have been conducted for any length of time. Prior to World War II the Navy's meager survey fleet did attempt systematic coverage of certain coastal areas. The war ended the attempt abruptly, and when hostilities ceased, the complexion of surveying changed radically. The priority of surveying was dictated by two considerations still valid today—the political nature of world affairs which involved United States peacekeeping efforts throughout the world and, second, recognition of the importance of oceanography to the nation and to naval doctrine. The necessity to increase surveys of all kinds and on a worldwide basis stemmed from these causes.

Multipurpose collection units

Since the magnitude of the subsequent effort exceeded available ship forces, the concept of multipurpose collection units evolved. Responsible naval officials held that, because ships were scarce and highly expensive to operate, each one should take advantage of every operation by obtaining the greatest amount of all kinds of data possible. Use of multipurpose ships did allow large amounts of all data to be collected, including bathymetric data, although the concept did not prevent assignment for collection of bathymetric data exclusively on particular missions.

Multipurpose usage led inevitably to high cost of new construction when, in 1960, Congress approved building new ships for surveying and oceanographic purposes. Before this time almost all such ships were conversions from other types. Although construction planning was based on optimum capability for a given cost, eventual costs exceeded first estimates by 10 to 30 percent. Perhaps it is time to reappraise the concept of multimission platforms: to specialize; to build smaller, less expensive, faster, single-purpose ships for bathymetric surveys; and to have ships available that can be dispersed to many areas or assembled in a concerted effort. Such a force could complete an assignment the first time and produce results lasting for many years without the need to schedule repeat operations. More and more specialized platforms will be developed for every aspect of oceanography; this is occuring even now.

Survey and oceanographic ships

The distinction between survey and oceanographic ships can be defined more clearly. A survey ship may be considered one that conducts hydrographic and bathymetric surveys primarily, limiting its oceanographic work to bathythermograph collection, if done at all, and sea state and current data. An oceanographic ship is virtually the converse, with the possible exception that it may conduct survey operations in limited areas to supplement its primary oceanographic endeavors. Both ships usually operate echo sounders in transit to and from operating areas. By these definitions, only the Navy ships conduct surveys. However, the Navy also has oceanographic ships, complicated further by their designation as "oceanographic survey" ships. The present rate of bathymetric data acquisition is theoretically not limited by the number of survey ships in use alone; this can be increased by utilization of both survey and oceanographic ships.

There are far greater numbers of oceanographic ships in operation than survey ships, even including in the latter group those nonoceanographic ships that can and sometimes do perform survey work. This last category includes several types—ones whose missions are related only indirectly to charting—cable layers, NASA range ships, gravity collection and other research ships. In general, they make excellent survey platforms, and this is true of many oceanographic ships. Somewhat under 100 oceanographic ships are in operation, of which perhaps one-half are of a size and so equipped as to conduct surveys. Since the ratio of even this number to the survey ships in operation is about 5:1, conversion of a relatively few oceanographic ships would greatly increase survey capabilities. Whether this is desirable, in consideration of the number and variety of existing oceanographic requirements, is a different problem. However, if an increase in the survey fleet became necessary, it could be accomplished relatively quickly.

Ships as survey platforms have been and are today largely indispensable, but their continued use must always be examined in light of numerous factors. Some of these are cost, ability to acquire the data through alternate means, technological innovations, etc. No system that would replace survey ships can now be foreseen, and it is safe to predict that ships will never be entirely replaced for survey work. But other collection means will be developed. A single platform for complete replacement of ships, however, is doubtful.

Helicopters as survey platforms

In theory, the use of helicopters as survey platforms is entirely feasible. Transducers can be towed beneath the surface (but not too deeply) connected electrically to the helicopter, and geographic control would be maintained by routine means. The concept has much to commend it, particularly speed of operation. Technologically there is no overwhelming reason why this kind of survey operation cannot be performed. But practical difficulties could offset the increased data acquistion rate. Helicopters can remain on station only a short time. Thus, extended over-water helicopter operations are impossible without ship bases. Data reduction facilities are minimal or nonexistent. And the economics of such operations, involving many helicopters and large support ships, has yet to be demonstrated. Development of longer range helicopters could change this forecast, but only if their in-flight endurance was considerably increased. Speed is of slight consideration, i.e., a speed of 50 to 100 knots is such a vast improvement over present survey ship speeds that further increase to 200 knots would have little bearing. Maintenance of given altitude above the ocean surface would be accomplished by use of upward-scanning sonar so true surface-to-bottom readings could be computed. In any event, helicopter surveys cannot be considered to have great potential in bathymetric work. Their potential for hydrographic work cannot, however, be as easily dismissed. Hydrographic work in its present form cannot be efficiently continued much longer.

Satellite platform

It is difficult to forecast lack of success for any sensing phenomenon associated with satellites. With regard to numerous oceanographic variables it is reasonably certain that satellites will eventually perform extremely useful functions. Sea state and ice conditions, surface current data, surface temperature, and other oceanographic information can and will be observed by satellites. Even inshore hydrography, reservoir sedimentation, glacial flow, and harbor survey work will benefit through satellite photography. But neither visual nor radar photographic techniques have any great potential for deeper bathymetric work. Unless spectacular breakthroughs in photography or laser technology occur, application to bathymetry cannot be predicted in the immediate future for satellite or other airborne platforms.

Submarine platforms

On first consideration, the ability to conduct survey operations while completely submerged offers a distinct advantage, because of the calm waters in which the platform operates. Surface waters are rough and can sometimes delay or halt survey operations. Submerged sounding is a reality now from the standpoint of putting sensors beneath the surface and through the use of completely submerged platforms, such as submarines and deep submergence vehicles (DSVs). Surveys by DSVs would probably be conducted only in limited areas. The drawback of submerged platforms for routine survey work is not lack of means to determine position, but the fact that surface ship operation is less expensive. Submerged platforms will be built to include bottom-survey ability, but none will be built for that purpose exclusively. Surveys will be conducted to augment the primary intended mission of the submerged vehicle. Positioning of vehicles close to the bottom remains a problem, but control near the surface can be obtained by towed whip antennas. Eventually, vehicles, utilizing fixed beacons for control, will conduct some surveys close to the bottom. Yet if a forecast were to be made, the probability of their use (or that of submarines) for routine bathymetric surveys is marginal.

Towed sensors, cable-controlled vehicles, and other platforms

Quiescence under the surface can be utilized by surface ships, towing submerged sensors, which eliminate positioning difficulties. Towed sensors will not be used for routine bathymetric collection on a large scale. Their chief use will continue to be to provide data for microbathymetric studies in relatively limited areas. Inshore hydrography may benefit from applications of paravane techniques. Remote control vehicles will be developed to transmit depth data in "real time," coupled with positioning and computerized control applications. Higher survey speeds will be available with hydrofoil and ground effect-type platforms. Man-in-the-sea programs may provide for "ground" observations of fixed-bottom control points; cable-controlled vehicles for salvage, survey work, and mining will be given increased range and utility.

Collection techniques

The data collection techniques of the future will employ the swath-sounding approach, side-looking sonar, underwater photography, lasers and holography, computer modeling, and control buoys. Some of these techniques are already in use; others may be researched and developed for the next ten years before they are used at sea.

Swath-sounding approach

One of the most impressive survey techniques developed during the past decade is *swath sounding*. Were it not for current limitation in its use, the concept might be termed revolutionary. Swath sounding would not have become an operational survey system without parallel advances in electronics and computer technology, and even so, development was slow and achieved only with great difficulty. Increased numbers of such systems in the future must imply a willingness to pay for the benefits derived, because of large development costs involved. This stems from miniaturization requirements for installation on smaller hulls, as well as for computer support.

It is believed that this miniaturization will occur and that additional swath-sounding units will be in use. Return in bathymetric data, whether determined by total program investment, cost per sounding mile, or cost per day at sea, dictates such a course. Procurement of individual units on a haphazard basis would be a mistake, however, without integration of hull characteristics, data-reduction techniques, computer application, and overall requirements. Further, techniques for translation of area coverage needs into survey specifications must be developed. This kind of development is particularly amenable to a typical systems approach, which could completely evolve new survey concepts and techniques.

Side-looking sonar

Side-looking sonar deserves technical improvement for supplemental survey purpose. This treatment places the axis of the transducer horizontally, allowing the acoustic rays to transmit and receive across, rather than from above, the ocean bottom. The advantage in this application is due primarily to the large bottom area covered. Vertical coverage is sufficiently maintained so that the bottom is "sounded" from an area very close to or even under the ship out to equipment ranges. To date, side-looking sonar can be used only in shallow waters or towed close to the bottom when in deeper areas. In either case a good definition of microbathymetry is achieved. Typical ranges depend on water conditions, target strengths, etc., but rarely exceed several miles or about five to six times the depth. The present configuration does not provide unambiguous depths for charting purposes, but it may be possible that this limitation can be overcome.

Underwater photography

Underwater photography is invaluable for many purposes, but will lack general mapping potential except for limited bottom areas. Effective range from camera to bottom has increased since the first remote deep-sea photograph was taken by Lamont Geological Observatory in 1939, but haze still limits it to about 30 meters. Literally thousands of photographs would be necessary per square mile of ocean bottom in order to compile any but the largest scale maps even if stereoscopic overlap were not required. Other than data reduction problems caused by this volume of photographs, camera positioning poses serious difficulty. Data reduction and positioning problems could be solved, but no known lighting system will overcome the range limitation caused by haze. If haze could be eliminated (as does not seem likely), the potential for production of bathymetric charts is present.

Lasers and holography

The laser principle was developed by the American physicist Charles H. Townes in 1960. The laser produces an intense beam of coherent light of high directivity, with the green to green-blue laser penetrating the farthest in the oceans, or rather, being absorbed the least. An invention in 1947 by Dennis Gabor of the Imperial College of Science and Technology in London was waiting for discovery of the laser: this invention was holography which freezes in film the optical wave fronts of an object and preserves original depth and parallax. The laser is used as the light source which reconstitutes on film the light waves of the object; without the laser the hologram appears as a greyish picture with a fine network of lines.

In acoustic holography, illumination of objects is accomplished by sound energy instead of laser beams as in optical holography. The resultant hologram is recorded using several different techniques and is reconstituted by lasers. One problem is that of perspective—the three-dimensional effect of the optical hologram is due to parallax achieved by the viewer. Because this kind of hologram is recorded at optical wavelengths, resolution is acceptable when viewed by the human eye. In acoustic holography, however, wavelengths are larger. Parallax effects using normal vision do not yield acceptable resolution, and other means to observe reconstituted images must be developed. However, the hologram and the laser open a new dimension in oceanography even though present difficulties in this infant science are manifold. The echo sounder has replaced the lead line; perhaps acoustic holography will someday replace the echo sounder as we know it today.

Computer modeling

Much remains to be done in terrain simulation and statistical extension of mathematical models by computer programs. Probabilistic models should provide a means for computer simulation of a wide variety of geological processes under the oceans. In the earth sciences generally, probabilistic means aid the study of stratigraphy, sedimentation, paleontology, and geomorphology. Experimental computer modeling will increase as the basic data matrix necessary to inaugurate programs is more and more obtainable. The experimental simulation of "real-world" processes appears dependent on the computer orientation of individual oceanographers, who are motivated in part by the interests of the organization to which they are attached and the direction of their graduate training.

Computer generation of relief perspectives is feasible and should prove valuable in studies involving areas of limited horizontal extent. The importance of looking at relief in this way, other than the advantage it affords in "reading" shape, lies in the various angles from which the perspective can be displayed. Figure 67 shows the computer-generated relief perspectives.

Control buoys

Lessons learned in position maintenance from deep-drilling operations will benefit survey work through development of automatic control buoys for special search and salvage. These buoys would overcome deep-anchoring problems and would contain thrusters which correct for deviation from a radius of position error.

Data reduction and libraries

Small-scale chart coverage, backlogged data, and repository of bathymetric data are considered under data reduction and libraries.

Small-scale chart coverage

Small-scale coverage of ocean bathymetry exists in several chart series. One of these is the General Bathymetric Chart of the Oceans (GEBCO) at a scale of 1:10,000,000. The International Hydrographic Bureau (IHB) initiated this program about 25 years ago, following earlier efforts by Prince Albert of Monaco. Seventeen member nations of the IHB compile and maintain GEBCO, which is published by the Institute Geographic National in Paris. World coverage is as yet incomplete.

The Bathymetric Chart (BC) series is published by the United States Naval Oceanographic Office. Despite its rather small-scale (but much larger than GEBCO) and overly large contour interval (100 fathoms), it can be and is used by submarines for submerged position location. Positions so achieved have accuracies of 1,000 yards in exceptional cases, but this is possible on the BC chart chiefly in areas compiled from controlled surveys. As a bathymetric chart suitable for general bathymetric navigation, the series does not suffice.

Project SEAMAP is an effort begun in 1961 by the United States Coast and Geodetic Survey and since hampered by shortage of funds and ships. Although bathymetric data were collected during these surveys, this was through spacing of 10-mile lines. Much excellent bathymetric data were produced as a result of the survey effort. To date, charts of areas in the North Pacific and off the coast of Southern California have been published. Lack of more detailed survey lines notwithstanding, world coverage even at 10-mile spacing can yield much worthwhile data.

Backlogged data

Occasional statements on 20-year backlogs of unprocessed survey data, or about 50 or 100 ship-years of survey requirements that must be met, actually make little sense and upon analysis are found to be misleading. Whether very great importance can be placed on the value of bathymetric backlogs is questionable.

Most of the information in this category consists of poorly controlled and random ship tracks. Although age in itself is not a hindrance, some of the data are very old. Moreover, the majority are not entirely unprocessed, as they received some review upon receipt for possible contribution to then-current production programs. The backlogged data should not form the major basis for chart production of certain areas, except with special care, even in areas where recorded depths are sparse. Means to scan and store the backlogged data are easily developed, but elimination of the condition should more properly be attacked by solving the present rate of influx, through shipboard data reduction or procedures that store incoming data automatically and selectively.

The phrase "survey requirements" is without meaning unless qualified in use. Identification of a 50-year program of survey requirements presupposes that specific chart series have been determined to be needed. Procedures for translation of chart needs to survey times, accuracies, instruments, and even platforms are lacking at this time and are not easy to develop. Witness the use of the term ship-years; does survey time equate to ship-years? What is a ship-year? Are linear lines or square miles used as a criterion? Do different type ships have different ship-years? These questions are not meant to be facetious, and illustrate the difficulties in quantifying requirements.

Repository of bathymetric data

A central repository of bathymetric information ranks high in priority of need. This can be patterned after similar sorting and library organization which have proven successful—World Data Centers and the National Oceanographic Data Center (NODC), for example. NODC does not include bathymetric information, partly because of shortages in personnel, space, and equipment, and in some measure to the belief of member organizations that other oceanographic activities should take precedence at this time.

In January, 1969, the Commission on Marine Science, Engineering, and Resources, issued a report entitled *Our Nation and The Sea* (Government Printing Office, Washington, D.C.), which recommends a National Oceanic and Atmospheric Agency (NOAA) to report directly to the President and to combine the present offices of the Bureau of Commercial Fisheries, U.S. Coast Guard, Environmental Science Service Administration, U.S. Lake Survey, National Sea Grant Program of the National Science Foundation, and the National Oceanographic Data Center. Over 300 seagoing ships would become available to NOAA, whose first mission would be systematic mapping of bathymetry of United States nearshore waters.

Other developments

Navigational scale, control systems, positional requirements, and an ideal bathymetric navigational system that can be developed are discussed below.

Navigational scale

Investigations are proceeding to determine the optimum chart for bathymetric navigation use. It is not difficult to forecast that an eventual product of this kind will be of larger scale than now available and will have a larger (closer) contour interval. No single scale need be decided upon covering all areas, and a 1:300,000-scale chart over certain topography might be appropriate, whereas a larger scale might be desirable over flatter areas. For ease of survey and chart construction, particular organizations may decide upon a single-scale series. An ideal scale might be supposed to be about 1:200,000, perhaps slightly larger. Scale is of less importance than utility, and this means accurate well-controlled portrayal of bathymetry.

Control systems

Control systems will be developed which will allow accuracies of ±500 yards in all areas of the world. These may be satellites, assuming that a sufficient number can be orbited and data reduction equipment becomes less sophisticated and costly. No satellite system will replace electronic control completely. Loran-C/D will replace Loran-A in five to ten years and will be extended to cover all but the remotest areas of the earth. Fixed underwater beacons will be increasingly utilized, but will not become the general geodetic panacea once envisioned. The most urgent requirement they will meet will be for submerged vehicles. However, long-lived reliable beacons could become important aids to pinpoint positioning on continental shelves. Their use in conjunction with bathymetric methods could aid survey, farming, and mineral exploitation.

Positional requirements

To reflect on requirements, the author contemplates that more precise positioning will be demanded in the future. However, the subject of requirements always promises to cause difficulty in achieving common viewpoints. It is not very difficult to achieve most of the positioning accuracies required by navigators today. More effort should be spent on forecasting the probable effect of future developments on the navigator's requirements many years hence. It is safe to predict such increased accuracies as being needed for many activities—farming, mineral exploitation, and so on. A real contribution to navigation would result if individuals could quantitatively determine postioning needs of the future. Systems subsequently developed could specify satisfaction of particular needs, rather than the needs accommodate themselves to the accuracies thereby available.

Ideal bathymetric navigational system

As one might presently envision an ideal navigational system, it would include console arrangements on or near the bridge to display the bathymetry of an area as the ship is in transit. One console would show gross bathymetry at a general navigation scale, indicating by means of a line or other CRT display mode the ship's track made good, present position, and DR advance. A large-scale console should reveal more detailed bathymetry, against the background of which the ship's position would be marked. Compact computers would correlate preprogrammed reference input with that observed in a continued series of updating events. These would analyze, accept or reject, and supplement observed information to feed instructions to a counter-viewer arrangement which would indicate actual position by latitude and longitude in preset time increments. Metric sounding data would be used.

The survey instrumentation could be a system of standard, array, directional, swath and (range and depth improvements permitting) side-scan and doppler mechanisms. The ship's course would be programmed earlier into a control computer, based upon topography, wind and sea conditions, currents, and late weather information. A decision to deviate from the planned transit would override the programmed courses, with means to return to, or determination of, a new transit plan.

What has just been described is merely one set of ideas and is not by any means the only or the best possible. This particular system is within current technical capability to develop. Costs aside, the decision to undertake such development is another matter, involving management considerations and prerogatives. With no little amount of temerity, future development of systems conceptually similar to that described is predicted.

Management of bathymetric programs

Manpower, administration, federal support, and the voice of the navigator are needed in the management of bathymetric programs.

Manpower

In the field of education and training, the urgent need for individuals trained in the ocean sciences has, of course, been recognized for some time. At present, there is also need for greater numbers of nonacademic persons—technicians—for engineering work primarily. Currently, programs in marine science are increasing at a rate greater than they can be accommodated with qualified manpower, and this situation will always be present to some extent. However, this is one problem which is receiving attention, with groups of competent individuals studying its solution.

Administration

In the administration of bathymetric collection programs, there is a danger in overspecialization. A nationwide shortage of administrative talent is conceded to be a continuing difficulty, and this is more notable in evolving disciplines. In government agencies, this shortage is acute at all levels, depending on individual organizational structures. Administration has recently been given renewed interest through the need for well-experienced personnel, whose awareness of issues cuts across technical specialties and agency lines. Many men in administrative positions today have been too long involved in specialized work, and as a result they have been given responsibilities for which their previous training has not equipped them. The administrator cannot remain a specialist if he is to be successful; rather he must broaden his abilities to consider his and other specialties as elements contributing to a whole discipline. Decision-making processes are sometimes the subject of derisive comment, with Department of Defense programs bearing the brunt of such comment. Yet proper decision-making is the keynote of the success or failure of many programs. Optimum development of programs in bathymetry will be facilitated if administrators have the perspective to make enlightened decisions.

Federal support

Federal financial support available to the marine sciences in the broadest sense will increase in the future, but it would be a mistake with serious implications to assume that government funds are unlimited or, indeed, that the present rate of increased spending will be maintained. The need for long-range planning cannot be disparaged; yet, experience has shown the fallacy of emphasizing exotic goals which are countless in oceanographic research, at the expense of more realistic and immediate objectives. There will be ample federal spending, but no spectacular spending splurge such as seen elsewhere. This very limitation will weed out efforts that are uneconomic and inopportune and will strengthen those programs which are. Comparison of oceanography with the national space effort is sometimes overstated, but it is believed valid to state that such comparisons have shown that programs must be realistic, well-planned, capable of achievement at a pace not to be accelerated beyond a certain point without serious consequences, and subject to re-evaluation and reconsideration consistent with constant fluctuation in requirements, schedules, technology, and resource availability. These same requirements must also apply to oceanography.

The Navigator's voice

The pace of advancement in all facets of oceanography is a rapid one. Information contained in this book may be outdated within the next few years. This is not a fault, but a situation that should be welcomed. Advances in electronics and optics alone must give material impetus to surveys, chart production, and navigational procedures which have heretofore been based on long-instituted concepts. Improvements benefiting the navigator will occur, whether he is directly involved in these developments or not and whether such improvements are considered specifically or as fallout from other primary efforts. To the extent that developments should or can proceed in response to recognizable needs, the navigator's requirements must be plainly stated and considered, and his voice heard early in the development cycle.

Appendix

CHART FROM BATHYMETRIC ATLAS

This area of relatively isolated seamounts is taken from an atlas, and therefore the scale (about 1:2,000,000) and contour interval (200 fathoms) make this particular product unsuitable for location to high accuracy. However, it shows a type of physiographic province which is ideal for navigation and positioning with the proper charts. One of several methods, e.g., Side-Echo or Profile-Matching, could be used with the bathymetry shown. Note that the seamounts are widely spaced and portray different overall shape and minimum depths. Greater credence might be placed on position of the features and their minimum depths, as their overly concentric shapes (at least at this scale) may be subject to question. This chart is taken from U. S. Naval Oceanographic Office's *Publication No. 1301, Bathymetric Atlas of the Northwestern Pacific Ocean, 1969.*

124   APPENDIX

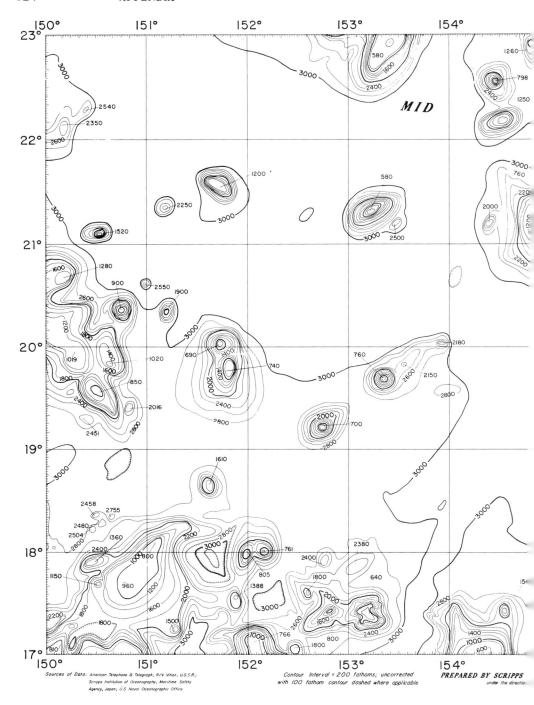

Appendix: Chart from bathymetric Atlas.

APPENDIX 125

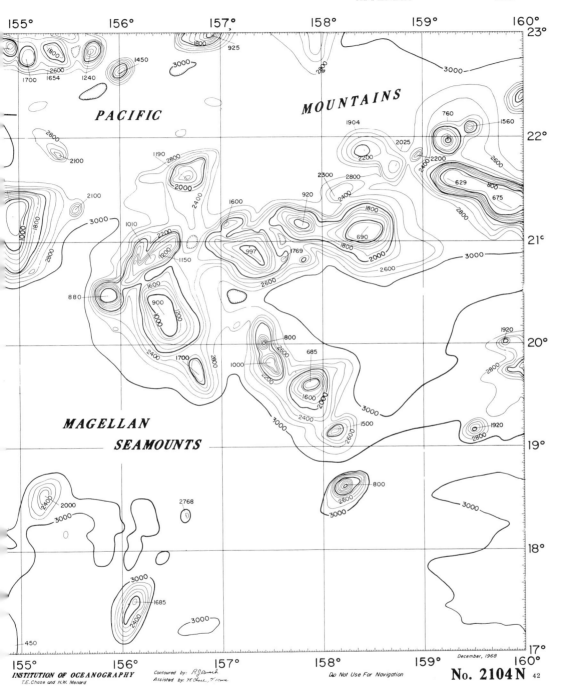

Glossary of Selected Terms

This glossary provides the navigator with selected terms applied to bathymetric navigation and charting. The definitions are taken from Special Publication No. 35 published in the fall of 1966 by the U.S. Naval Oceanographic Office and used herewith by permission of the Oceanographer of the Navy. It should be noted that some of the terms are defined through use of best judgment factors while others are authoritative in that they exist in some formalized way.

Absorption loss—That part of the transmission loss which is due to dissipation or the conversion of sound energy into some other form of energy, usually heat. This conversion may take place within the medium itself or upon a reflection at one of its boundaries.
Abyss—A particularly deep part of the ocean, or any part below 300 fathoms.
Abyssal (or abyssobenthic)—Pertaining to the great depths of the ocean, generally below 2,000 fathoms (3,700 meters).
Acoustic dispersion—The scattering or spreading of sound with frequency.
Acoustics—The science of sound, including its production, transmission, and effects.
Acoustic scattering—The irregular reflection, refraction, or diffraction of a sound in many directions.
Ambient noise—The noise produced in the sea by marine animals, ship and industrial activity, terrestrial movements, precipitation, and other underwater or surface activity outside the measuring platform and detection equipment itself.
Array—A group of two or more devices such as hydrophones which feed into a common receiver. The purpose of thus grouping hydrophones is to increase coverage and sensitivity of the listening unit and also to determine the bearing of a target.
Arrival—The chronologic appearance (such as first, second, third arrival) of different wave energies on a seismic or acoustic record.
Attenuation—The reduction in sound or light intensity caused by the absorption and scattering of sound or light energy in air or water.
Axis of acoustic symmetry—For many transducers the three-dimensional directivity is such that it may be represented by the surface generated by rotating a two-dimensional directivity pattern about the axis corresponding to the reference bearing of the transducer. This axis may then be described as an axis of acoustic symmetry, or, more briefly, as the acoustic axis.

Backscattering—The part of the reflected sound energy that returns to the transducer; equivalent to reverberation.

Bank—1. An elevation of the sea floor located on a continental (or island) shelf and over which the depth of water is relatively shallow but sufficient for safe surface navigation. It may support shoals or bars on its surface which are dangerous to navigation.

2. In its secondary sense, a shallow area consisting of shifting forms of silt, sand, mud, gravel, etc., but in this case it is only used with a qualifying word such as sandbank, gravelbank, etc.

Bathymetric chart—A map delineating the form of the bottom of a body of water, usually by means of depth contours (isobaths).

Bathymetry—The science of measuring ocean depths in order to determine the sea floor topography.

Beam width—The beam width of a directional transducer of a given frequency in a given plane which includes the beam axis is the angle included between the two directions, one to the left and the other to the right of the axis, at which the angular deviation loss has a specified value.

Bottom reflection—The return of transmitted sound from the bottom of the ocean. The characteristics of reflected sound depend on the nature of the bottom and on the wavelength of the sound.

Canyon—A relatively narrow, deep depression with steep slopes, the bottom of which grades continuously downward.

Cavitation—The turbulent formation, generally mechanically induced, including growth and collapse of bubbles in a fluid, and occuring when the static pressure at any point in fluid flow is less than fluid vapor pressure.

Cavitation noise—The noise produced in a liquid by the collapse of bubbles that have been created by cavitation.

Continental margin—A zone separating the emergent continents from the deep-sea bottom; generally consists of the continental shelf, slope, and rise.

Continental rise—A gentle slope with a generally smooth surface, rising toward the foot of the continental slope.

Continental shelf (also called continental platform)—A zone adjacent to a continent or around an island, and extending from the low-water line to the depth at which there is usually a marked increase of slope to greater depth.

Continental slope—A declivity seaward from a shelf edge into greater depth.

Contour—A line on a chart representing points of equal value with relation to datum. It is called an isobath when connecting points of equal depth below sea level.

Contour interval—The difference in value between two adjacent contours.

Cordillera—An entire mountain system, including all the subordinate ranges, interior plateaus, and basins.

Decibel (abbreviated db)—A value that expresses the comparison of sound of two different intensities. This value is defined as 10 times the common logarithm of the ratio of the two sound intensities.

Deep scattering layer (also called DSL, false bottom, phantom bottom)—The stratified population(s) of organisms in most oceanic waters which scatter sound. The scattered sound is recorded on echo sounder as a uniform, horizontal band or stripe, and such layers generally are found during the day at depths from 100 to 400 fathoms. A layer rarely is less than 25 fathoms thick and may be as much as 100 fathoms thick. Several layers often are recorded at the same time and may be continuous horizontally for many miles. Most layers typically undergo diurnal vertical movements.

Depth—The vertical distance from a specified sea level to the sea floor. The charted depth is the recorded distance from the tidal datum to the bottom surface at the point, using an assumed velocity of sound in water of 800 fathoms per second (U.S.) and with no velocity or slope corrections made.

GLOSSARY OF SELECTED TERMS 129

Direction-response pattern (or beam pattern)—The directional-response pattern of a transducer used for sound emission or reception is a description, usually presented graphically, of the response of the transducer as a function of the direction of the transmitted or incident sound waves in a specified plane and at a specified frequency. In general, the beam pattern will change with a change in the operating frequency.

Directivity—The confining of sound to a beam by mechanical and/or electronic means.

Directivity index—A measure of sound pressure level in one direction compared to that in all other directions.

Dome—A transducer enclosure, usually streamlined, used with echo-ranging or listening devices to minimize turbulence and cavitation noises arising from the passage of the transducer through the water.

Echo—An acoustic signal which has been reflected or otherwise returned with sufficient magnitude and time delay to be detected as a signal distinct from that directly transmitted.

Echogram—The graphic presentation of echo soundings recorded as a continuous profile of the bottom.

Echo ranging—The determination of distance by measuring the time interval between transmission of a radiant energy signal (sound) and the return of its echo.

Echo sounding (or acoustic sounding)—Determination of the depth of water by measuring the time interval between emission of a sonic or ultrasonic signal and the return of its echo from the bottom. The instrument used for this purpose is called an echo sounder.

Electroacoustic transducer—A transducer for receiving waves from an electric system and delivering waves to an acoustic system, or vice versa.

Electromechanical transducer—A transducer for receiving waves from an electric system and delivering waves to a mechanical system, or vice versa.

Electrostatic transducer—A transducer that consists of a capacitor and depends upon interaction between its electric field and the change of its electrostatic capacitance.

Escarpment (or fault scarp, scarp)—An elongated and comparatively steep slope of the sea floor, separating flat or gently sloping areas.

Fathom—The common unit of depth in the ocean for countries using the English system of units, equal to 6 feet (1.83 meters). It is also sometimes used in expressing horizontal distances, in which case 120 fathoms make one cable or very nearly one-tenth nautical mile.

Fathometer—Copyrighted trade name for a type of echo sounder. Often incorrectly used for any sonic submarine depth measuring system.

Fault coast (or fault-scarp coast)—A straight coast formed by a fault, consisting of a seaward facing escarpment and a downthrown block below sea level.

Fish—Any towed sensing device.

Fix—A relatively accurate position determined without reference to any former position. It may be classed as visual, sonic, celestial, electronic, radio, hyperbolic, loran, radar, etc., depending upon the means of establishing it. A pinpoint is a very accurate fix, usually established by passing directly over or near an aid to navigation or a landmark of small area.

Frequency of sound—The number of sound waves passing a point in a given time; usually measured as cycles per second.

Hachures—1. Short lines on topographic or nautical charts to indicate the slope of the ground or submarine bottom. They usually follow the direction of the slope.
2. Inward-pointing short lines or "ticks" around the circumference of a closed contour indicating a depression or a minimum.

Hydrodynamic noise—Noise produced by the motion of the ship or sonar housing through the water.

Hydrographic survey—A survey of a water area, with particular reference to submarine relief, and of any adjacent land.

Hydrography—The science which deals with the measurement and description of the physical features of the oceans, seas, lakes, rivers, and their adjoining coastal areas, with particular reference to their use for navigational purposes.

Hydrophone—An electroacoustic transducer that responds to water-borne sound waves and delivers essentially equivalent electric waves.

Interface (also called internal boundary, surface of discontinuity, or boundary surface)—A surface separating two media, across which there is a discontinuity of some property, such as density, velocity, etc., or of some derivative of one of these properties in a direction normal to the interface.

Isobath (sometimes called fathom curve, depth contour, and depth curve)—A contour line connecting points of equal water depths on a chart.

Knoll—An elevation rising less than 500 fathoms (1,000 meters) from the sea floor with limited extent across the summit.

Layer depth effect—The weakening of the sound beam owing to abnormal spreading as it passes from an isothermal or a positive gradient layer to an underlying negative layer.

Lead line (or sounding line)—A line, wire, or cord used in sounding. It is weighted at one end with a plummet (sounding lead).

Line hydrophone—A directional hydrophone consisting of a single, straight-line element, or any array of contiguous or spaced electroacoustic transducing elements, disposed on a straight line, or the acoustic equivalent of such an array.

Lobes—If a three-dimensional representation of a transducer directivity is made by rotating the two-dimensional directivity pattern these sectors generate zones, or regions, on the constant distance surface. These regions are known as lobes. The term is also used with reference to the corresponding portions of the directivity pattern. The region, or sector, which includes the reference axis is known as the primary lobe; the remaining regions, or sectors, are known as the secondary lobes.

Map—A representation on a plane surface, at an established scale, of the physical features (natural, artifical, or both) of a part or the whole of the earth's surface, by means of signs and symbols, and with the means of orientation indicated.

Midocean ridge—A great median arch or sea-bottom swell extending the length of an ocean basin and roughly paralleling the continental margins.

No-bottom—A notation appearing on nautical charts indicating that the sounding did not reach the bottom.

Oceanographic survey—A study or examination of conditions in the ocean or any part of it, with reference to animal or plant life, chemical elements present, temperature gradients, etc.

Own-ship's noise (or self noise)—Often the limiting noise produced by the ship (or equipment) itself or as a result of its motion, and registered by a sonar receiver.

Passive sonar—A method or equipment by which information concerning a distant object underwater is obtained by evaluating the sound generated by the object itself.

Piezoelectric effect—The phenomenon, exhibited by certain crystals, in which mechanical compression produces a potential difference between opposite crystal faces, or in which an applied electric field produces corresponding changes in dimensions.

Province—A region composed of a group of similar bathymetric features whose characteristics are markedly in contrast with surrounding areas.

Reflection of sound—The process whereby a surface of discontinuity turns back a portion of the incident sound into the medium through which the sound approached.

Relief—The inequalities (elevations and depressions) of the sea bottom.

Reverberation—Sound scattered towards the source, principally from the ocean surface (surface reverberation) or bottom (bottom reverberation), and from small scattering sources in the medium such as bubbles of air and suspended solid matter (volume reverberation).

Reverberation index—The measure of the ability of an echo-ranging transducer to distinguish the desired echo from the reverberation. Computed from the directivity patterns as ratio in decibels of the bottom-, surface-, or volume-reverberation response of a specific transducer to the corresponding response of a nondirectional transducer.

Seachannel—A long, narrow, U-shaped or V-shaped shallow depression of the sea floor, usually occurring on a gently sloping plain or fan.

Seamount—An elevation rising 500 fathoms (1,000 meters) or more from the sea floor with limited extent across the summit.

Seamount chain—Several seamounts in a line with bases separated by a relatively flat sea floor.

Seamount group—Several closely spaced seamounts not in a line.

Seamount range—Several seamounts having connected bases and aligned along a ridge or rise.

Shoal—A submerged ridge, bank, or bar consisting of, or covered by, unconsolidated sediments (mud, sand, gravel) which is at or near enough to the water surface to constitute a danger to navigation. If composed of rock or coral, it is called a reef.

Slumping—The slow creeping of mud and other debris down a slope, caused by gravity.

Sonar—1. An acronym derived from the expression "*so*und *n*avigation *a*nd *r*anging." The method or the equipment for determining by underwater, sound techniques the presence, location, or nature of objects in the sea.

2. A system for determining distance of an underwater object by measuring the interval of time between transmission of an underwater sonic or ultrasonic signal and return of its echo.

Sound channel—The region in the water column where sound velocity first decreases to a minimum value with depth and then increases in value as a result of pressure. Above the minimum value, sound rays are bent downward, and below the minimum value, sound rays are bent upward; the rays are thus trapped in this channel. Sound traveling in a deep channel can be detected thousands of miles from the sound source.

Sound velocity—The rate of travel at which sound energy moves through a medium, usually expressed in feet per second.

The velocity of sound in seawater is a function of temperature, salinity, and the changes in pressure associated with changes in depth. An increase in any of these factors tends to increase the velocity. Sound is propagated at a speed of 4,742 feet per second at 32°F, one atmosphere of pressure, and a salinity of 35 per mille.

Subbottom reflection—The return of sound energy from a discontinuity in material below the sea-bottom surface.

Submarine geomorphology—The branch of geology that deals with the features of the sea floor, their form, origin, and development, and the changes they are undergoing.

Submarine well—A cavity on the sea bottom; also called a submarine pit.

Tableknoll—A knoll with a comparatively smooth, flat top.

Tablemount—A seamount having a comparatively smooth, flat top.

Taut-wire mooring—A mooring arrangement in which a submerged float provides the upward force necessary to maintain the system in a fixed position with reference to the sea bottom. Taut-wire moors may be single, double, or multipoint according to design requirements of the system and according to the speed and variability of the ambient currents.

Transducer—A device that converts electrical energy to sound energy, or the converse. When sound energy received through the water is converted to electrical energy, the device is termed a hydrophone; conversely, when electrical energy is converted to sound energy and transmitted through the water, the device is termed a sonar projector or echo sounder.

Trench—A long, narrow, and deep depression of the sea floor with relatively steep sides.

Trough—A long depression of the sea floor normally wider and shallower than a trench.

Turbidity current (or density current, mud flow, suspension current)—A highly turbid, relatively dense current carrying large quantities of clay, silt, and sand in suspension which flows down a submarine slope through less dense sea water.

Bibliography

Bowditch, N. 1962. *American practical navigator.* Washington: U. S. Naval Oceanographic Office.

Bagrow, L. 1964. *History of cartography.* ed. R. A. Skelton. London: C. A. Watts & Company Ltd.

Baker, B. B. Jr., Deebel, W. R., and Giesenderfer, R. D., eds. 1966. *Glossary of oceanographic terms.* Washington: U.S. Naval Oceanographic Office.

Clay, C. S., Ess, J., and Weissman, I. 1964. Lateral echo sounding of the ocean bottom on the continental rise, *Journal of Geophysial Research,* vol. 69, no. 18, September 1964.

Cohen, P. M. 1964. Bathymetric navigation. *U.S. Naval Institute Proceedings.* vol. 90, no. 10, pp. 66–71.

———. 1959. Directional echo sounding on hydrographic surveys. *International Hydrographic Review.* vol. XXXVI, no. 1. pp. 29–42.

Collinder, P. 1955. *A history of marine navigation.* New York: St. Martin's Press.

Commission on Marine Science, Engineering, and Resources 1969. *Our nation and the sea.* Washington: Government Printing Office.

Duncan, J., Captain, USN 1968. Personal communication in January 1968 concerning the use of bathymetry to position U.S.S. *Floyds Bay (AVP–40).*

Edvalson, F. M. 1965. *Sea floor names in principle and practice.* 10th Pan American Institute on Geography and History Symposium.

Environmental Sciences Service Administration 1966. *Development potential of U.S. continental shelves.* Washington.

Gilg, J. G., and McConnel, J. J., Jr., 1966. *Nonexistent seamounts.* Informal manuscript 66-28. U.S. Naval Oceanographic Office.

Greenhood, D., 1964. *Mapping.* Chicago: University of Chicago Press.

Heezon, B. C., Tharp, M., and Ewing, M. *The floors of the oceans.* Special Paper 65. New York: Geological Society of America.

Hewson, J. B. 1951. *A history of the practise of navigation.* Glasgow: Brown.

Karo, H. A. 1962. Hydrographic automatic data processing. Paper presented at 8th International Hydrographic Conference, Monaco.

Krause, D. C., and Menard, H. W. 1965. *Marine geology.* Depth distribution and bathymetric classification of some sea-floor profiles. pp. 170–179. New York: Elsevier Publishing Co.

Stephan, J. G. 1966. Mapping the ocean floor. *Battelle Technical Review.* July-August, 1968.

Taylor, E. G. 1957. *The haven-finding art.* New York: Abelard-Schuman.

U.S. Department of State. 1965. Sovereignty of the sea, *Geographic Bulletin,* no. 3. Washington.

U.S. President's Science Advisory Committee. 1966. *Report of President's Science Advisory Committee on effective use of the sea.* Washington.

Waters, D. W. 1958. *The art of navigation in England in Elizabethan and early Stuart times.* New Haven: Yale University Press.

Williams, W. M. 1968. Personal communication concerning statistical probabilities on use of contour advancing for position determination.

Index

Page numbers for definitions are in *italics*

Abrupt emergence of features, 46–47
Abyssal areas on ocean bottom
  abrupt boundary of, 48
  as position indicator, 76
  as shown on echogram, 49
Accuracy of chart, 74, 108
Acoustic holography, 115–116
Acoustic sounder, developer of, 13
American Practical Navigator, 4–5
American Scout Seamount, example of nonexistent feature, 25–26, 35
Army Topographic Command, 104
Automation
  early application of, 94
  of data analysis and chart production, 100
  of future positioning systems, 119.
  See also Computer applications

Bathymetric atlas, 123–125
Bathymetric chart, *50*
  classification of 53, 54
  series, 117
Bathymetric data
  backlogged, 117–118
  repository of, 118
  sources of, 100–101
Bathymetric programs, management of, 120–121
Bathymetry, *6*, 93
BC (Bathymetric Chart) series, 117
Beam width of echo sounders, 18, 20
Behm, Alexander, 13
Bottom features
  categories of design on echogram, 78
  comparison of, with land features, 36
  geological description of, 35
  physiographic diagram of, 38
Bottom profile
  composite nature of, 22
  vertical exaggeration of, 18, 28, 54
Bowditch, 4–5
Bureau of Commercial Fisheries, 118

Cable-controlled vehicles, as survey platforms, 114
Canyon, *37, 47*
  as positioning aid, 47–48, 78
  as shown on echogram, 48
Casa de la Contractacion de las Indias, 52
Cavitation, *28*
Chart coverage, small-scale, 17
Chart portrayal, 52–53
Chart production, application of automation to, 100–103
Charts, *51*
  accuracy criteria of, 108
  choice of, for positioning, 74
  classification of, 53
  horizontal scales of, 53, 59, 118
  storage of, in computer, 104
Choice of charts, 74
Coast and Geodetic Survey. *See* Systematic surveys
Coast Guard, U.S., 118
Collection techniques, 114–116
Columbus, Christopher, 52
Commission on Marine Science, Engineering, and Resources, 118
Computer applications
  and chart production, 100–103
  and data handling, 93–100
  for navigation, 103–105
  requirements for, 106
  to statistical extension of land form, 116.
  to surveying, 91
  See also Automation and Swath Sounding
Computer-generated relief drawing, 105, 116
Computer modeling, 116
Continental margin, *37. See also* Continental slope
Continental shelf, *36*
Continental slope, *37. See also* Bottom features
Continuous profiles, 14–16
Contouring of geologic features, 37–40, 57

135

Contours, 52
  difficulties in obtaining, 65
  interpretation during compilation of, 61
  intervals of, 59–65
  relation of, to shape, 59
Contour advancement, positioning technique, 84–86
Control
  beacons, 4
  buoys, 116
  future capabilities, 119
  relation of, to contour shape, 14, 16
  for survey work, 40, 57
Controlled charts. *See* Charts
Correctness of chart, 108
Criteria for satisfactory chart, 108–109

Dana, Richard Henry, 12
Data analysis, 100
Data handling, 93–100
Data reduction, 117–118
Deep scattering layer, 24
  as shown on echogram, 25
Deep trenches, 37
Depth-determining instrument, early, 12
Depth range, of echo sounders, 27
Determination of position, 75
Directional transducers, 19
Directivity. *See* Beam width

Echo sounder
  concurrent use of different types, 20
  depth range, 27
  early types of, 11, 13
  operational statistics of, 30
  types in use, 17, 19, 27
Echogram
  illustration of, 50
  interpretation, 33, 50.
  See also Recorder
Enlarged recorder, 30–33
Environment, ocean, 5
Environmental Science Services Administration (ESSA), 118
Escarpments, 45, 78

False bottom. *See* Deep scattering layer
Federal support to oceanography, 121
Fessenden, R. A., 13
Fixed transducers, 20

Gabor, Dennis, 116
GEBCO, 117
General bathymetric chart of the oceans (GEBCO), description of, 117

Geologic features, contouring of, 37–40.
Geological integrity, loss of, 62
Geological interpretations, 38–40
Geology of the oceans, 35–50
Gravimeter, 75
Gravity data, relation of positional accuracy to collection of, 75
Great Meteor Seamount, 35

Hanno, 10
Hayes, H. C., 13
Helicopters, as survey platforms, 112
Herodotus, 9–10
Hills, 37
Historia, 9–10
Holographic techniques, 116
Holography, acoustic, application of, to oceanography, 115–116
Hydrographic office, first established, 52
Hydrophone, 17
Hyperbolic lines of position, 58
Hyperbolic trace
  cause of, 22
  as shown on echogram, 23, 47

International Hydrographic Bureau (IHB).
  See General bathymetric chart of the oceans
Interpretations
  geological, 38–40
  random data, 65–68
  survey data, 69–72
Island arcs, 37

Kelvin-White sounding machine, 12

Langevin, Pierre, 13
Laser
  and holography, 115–116
  principle, developer of, 115
Lead line, description of, 11
Line of soundings
  automation of, 103
  as positioning technique, 87–88
Lows, 48–49

Magnetic tape, storage of sounding data on, 94–95, 98
Manpower needs, 120
Maps and charts, distinction between, 51, 110
Marti, 13
Master sheet, production of, 101

Mid-Atlantic Ridge, 37
Multiple-mode systems, 95
Multiple returns, *23*
　as shown on echogram, 24, 46
Multipurpose collection units, 111

National Aeronautics and Space Administration (NASA), positioning of range instrumentation ships, 4, 104
National Oceanic and Atmospheric Agency (NOAA), 118
National Oceanographic Data Center (NODC), 118
National Sea Grant Program of National Science Foundation, 118
Naval Oceanographic Office, 55, 100, 117
Navigation,
　bathymetric and space, 103–104
　comparison of, with precise location needs, 75
　computer applications for, 103–105
　positioning requirements for, 5
Navigational control, 57
Navigational scale, 118–119
Navigational system, ideal bathymetric, 119–120
Navigator's needs, 121
Nearchus, 10
Nondirectional transducers, 19

Oceanographic ships, 111–112
Opacity of the ocean environment, 5
Oscillator/receiver, developer of, 13

Periplus, 9
Phantom bottom. *See* Deep scattering layer
Pilotage, modern, 10–11
Plains, 37, 48–49
Plotting sheets, 101, 102
Portrayals, modified contour, 72–73
Positioning techniques, 76, 79, 87
Primary lines of survey sounding, 56
Prince Albert of Monaco, 117
Procedural techniques, 79–91
Profile matching
　automation of, 103
　positioning technique, 79
Project SEAMAP. *See* Systematic surveys
Projector, as function of transducer, 17
Punched cards, storage of sounding data on, 94–95, 98

Random sounding data
　comparison of, with survey data, 53
　interpretation of, 65
Recognition techniques, 76–79

Recorder
　description of, 27
　enlarged type, 30
　illustrations of, 31, 32
　and transducer, 17
Recovery of position, 75, 76
Reference contour, 86
Reliability of chart, 108, 109
Relief drawings, computer-generated, 105, 116
Remote-control vehicles, as possible survey platforms, 114
Requirements
　for bathymetric data, 105
　difficulty in establishment of, 107
　future positional, 119
　for precise positioning, 4
　for work in the sea, 108
　See also Navigation
Rises, 37
Rutters. *See* Sailing directions

Sailing directions
　comparison of contents with Rutters, 10
　U.S. Navy, 11
Satellites, as oceanographic survey platforms, 113
Scale
　of charts, 54
　choice of, for positioning, 74.
　of recorders, 30, 42
　relation of, to contour interval, 57
　See also Charts
Scylax of Caryanda, 9
Seaknolls, 37, 63
Seamount
　Great Meteor, 35
　as positioning aid, 4, 20, 78, 88
　as shown on echogram, 43, 46.
　See also Side-echo technique
Shallow features, 45
Ship's track on bathymetric chart, 50
Ships, differences between survey and oceanographic, 111
Side echoes, *22*
　as shown on echogram, 23
Side-echo technique, as positioning means, 88–91
Side-looking sonar
　description of, 115
　as possible aid in seamount detection, 90
Simultaneous sounding, 20–21
Slope effect, correction for, 55
Sonic sounding apparatus, developer of, 13
Sound velocity
　correction for, 55
　factors determining, 21
　profile of, 21
Soundings, *14*
　automated collection of, 94

early use of, 9–10
track direction for collection of, 56
See also Lead line
Spaced protuberances, 47
Spot depths, 14–16
difficulty in gaining shape from, 15
Stabilized transducers, 19–20
Standard Navy echo sounder, 28–30
Strabo, 10
Submarine canyon. See Canyon
Submarine geology. See Bottom features
Supersonic echo sounder, developer of, 13
Supplemental lines of survey sounding, 56
Survey platforms, 111–114
Survey ships, 111–112
Swath sounding, 95–98
as automated technique, 94
comparison of, with other methods, 98
described, 95
future use of, 114–115
Survey techniques, described, 54, 57
Systematic surveys
prior to World War II, 111
Project SEAMAP, 74, 117
Submarine as survey platform, 113–114

Tableknolls, 37
Terrain-accuracy standards, 109–111
Terrain interpretation, 61–65
Terrain simulation. See Computer applications

Thompson, Sir William. See Kelvin-White sounding machine
Towed sensors as survey platforms, 114
Towed transducer, 18
Townes, Charles H., 115
Transducer
description of, 17, 30
directional, 19
fixed, 20
nondirectional, 19
stabilized, 19
towed, 18, 28
See also Beam width
Transmitter, keying and gating of, 31, 42

Uncontrolled charts. See Charts
Underwater photography, limitations of, 115

Vertical exaggeration. See Bottom profile

Weekly Notice to Mariners, 11
Wide sound cone, display effects of, 79
Work/tasks and associated terrain-accuracy standards, 109–111
World Data Centers, 118
WWV, radio station, 33

*Composed in ten-point Times Roman leaded three points
by Monotype Composition Company, Baltimore, Maryland*

*Printed by offset on sixty-pound Clear Spring Book Offset,
Antique White by Westvaco*

The cloth covering material is Holliston Roxite C, Vellum Finish

*Presswork by Universal Lithographers, Inc.
Cockeysville, Maryland*

Bound by L.H. Jenkins, Richmond, Virginia

Line drawings by WILLIAM J. CLIPSON

Designed by HARVEY SATENSTEIN

Edited by KENNETH G. WALSH, JR.